Astronomers' Universe

Other titles in this series

Calibrating the Cosmos: How Cosmology Explains Our Big Bang Universe
Frank Levin

Origins: How the Planets, Stars, Galaxies, and the Universe Began
Steve Eales

Jack Meadows

The Future of the Universe

A.J. Meadows
Department of Information Studies
Loughborough University
Loughborough
Leics. LE11 3TU
UK
jack.meadows@btinternet.com

The plates in this book are reproduced courtesy of the U.S. National Aeronautics and Space Administration (NASA).

Library of Congress Control Number: 2006921441

ISBN-10: 1-85233-946-2
ISBN-13: 978-1-85233-946-3

Printed on acid-free paper.

9 8 7 6 5 4 3 2 1

Springer Science+Business Media
springer.com

/08

Contents

Preface

Many excellent books have been written about the past history of the universe and of the various objects—galaxies, stars, planets—to be found in it. All the exciting events from the original Big Bang to the appearance of human beings have been carefully recorded. Much less has been written about what comes next. What will happen to all these galaxies, stars, and planets in the future? And what will happen to us, and to any other intelligent life in the universe? It is obviously more difficult to examine the future than the past, but there are ways of doing it. Not everything in the universe is the same age; so a study of the older objects gives us some idea of what will happen to the younger objects. Some things vary in a fairly regular way, so you can guess what will happen next. For example, the number of spots visible on the Sun's surface increases and then decreases again every eleven years on average. These ups and downs can be expected to continue for a considerable time in the future. Finally, theoretical explanations of how things work at present often give some hint of how they will develop in the future.

One rule-of-thumb in astronomy—though there are plenty of exceptions—is that the further away objects are, the less we know about them. This means that it is often more difficult to forecast the future for distant objects than for ones nearby. For this reason, the early chapters of the book start with the solar system (especially the Earth) because we know most about it. The latter part of the book ventures out further into space to discuss the future of galaxies and the universe. Of course, what happens in nearby space is affected by happenings elsewhere: the future of the universe may well influence the future of the solar system. So some themes—life in the universe, for example—come up for mention in more than one chapter.

We must never forget that science is always in a state of flux. Ideas about the world around us have changed greatly over the past few decades; we must surely expect further significant changes over the next few decades. Some knowledge—for example, about how stars evolve—seems firmly based. Other knowledge—for example, about the expansion of the universe—seems to change almost from month to month. So the vision of the future presented here should simply be taken as a best guess based on current evidence.

To Begin With . . .

Forecasting the future—if only we could! What happened in the past is interesting, but the really fascinating thing is what may happen in the future. The problem, of course, is that we know a fair amount about the past, but little about the future. Yet we plan our lives on the basis of guesses about what will happen next. Shall we have a walk tomorrow? It depends on the weather forecast—that is, the best guess that meteorologists can make about tomorrow's weather. Such guesses can be more or less likely. Will the Sun rise tomorrow? How many people will answer "no" to that question? Will my local football team win on Saturday? Well, it all depends: they may have a reasonable chance. The more complicated the event, the less easy it is to guess what will happen. But also, the more far away the event in time, the more difficult it is to guess what will happen. Suppose I ask: will the Sun rise on a morning a hundred million years from now? Are you as confident in your reply as you are for a sunrise tomorrow?

What we are continually doing is using our knowledge of the past and present to try and predict what will happen in the future. Meteorologists collect today's weather data and look at past weather patterns to try and decide what tomorrow's weather will be like. I look at the recent record of my local football club, plus such things as current injuries to players, to try and guess what will happen in their next match. How good my guess will be depends on how good the information I have is, and how well I am able to use it for prediction. This is as true for predicting the future of the universe as it is for predicting weather or football results. In this book, guesses about the future are based on what most scientists currently regard as our best understanding of the past and present of the universe. There are two words here that need looking at. The first is "most." Most scientists are in agreement about the current picture of the world, but there are always a few mavericks who disagree with one aspect or another of the picture. Usually, such mavericks are proved, in due course, to be wrong. But occasionally they are proved to be right, and that means that the generally accepted picture has to be changed. The second word is "current." Because science is continually changing, the picture that scientists had of the world a hundred years ago was considerably different from the picture we accept today. Correspondingly, we can expect that scientists a hundred years in the future will have a picture that differs appreciably from ours.

It might seem that we are backed into a corner here. If the scientific picture is going to change, why should any present-day forecast of the universe's future be regarded as worthwhile? The answer—as the rest of this book will make clear—is that some parts of the picture have a good solid basis, while others are more speculative. Correspondingly, some predictions are not likely to change drastically, while others may change this year, next year, or sometime. If one way of understanding the universe has worked satisfactorily for many years, then we are more likely to feel confidence in its predictions for the future. For example, Newton's ideas on gravitation and motion have been around for over three centuries. They are used all the time for predicting where the Moon and the planets will be in the sky at some date in the future. As anyone knows who has used these predictions to point a telescope, they are always spot on. But, even here, developments of recent years have thrown up a query. It has been found that complicated systems—and the solar system, with all its various bodies, is certainly that—can sometimes develop in unexpected, chaotic ways. This will not affect predictions of where the planets will be next year, but the further you look into the future, the more difficult it becomes to be certain what will happen. Ask how the planets will circle the Sun a thousand million years from now, and the answer will be mainly guesswork. So there are two reasons why predictions of the future might go astray. The first, and commoner one, is because we do not know enough about what is happening. The second is because predictions can get a bit fuzzy if you try to look too far into the future.

Us and Them

One way forward is to forget about what we know at present and to speculate along the lines of "what would happen if?" This is the science-fiction route. Occasionally, it comes up with surprising results. Jonathan Swift wrote *Gulliver's travels* in the early eighteenth century. It is actually a work of science fiction, although not usually classified that way. Gulliver visits five different countries, one of which is inhabited solely by mad scientists. (Yes, they had them in those days.) One of their discoveries is that Mars has two moons circling it in orbits that they have carefully determined. In reality, the two Martian moons were not discovered until a century and a half later, when they were found to have orbits not too different from those reported by Swift's astronomers. How did he do it? Well, it was known in his day that Venus had no moons, the Earth had one, and Jupiter had four. So it was speculated—and not only by Swift—that the further planets were from the Sun, the more moons they should have. To fit into this sequence, it was speculated that Mars should have two moons. But it is still a matter of debate why he chose those particular orbits.

Swift's guess was unusually good. Where it is possible to test science-fiction predictions, they are wrong far more often than they are right. In the nineteenth century, Jules Verne wrote a widely read story about a manned space capsule from the Earth orbiting the Moon. Sounds like a good prediction? The problem is that many of his assumptions are questionable. For example, his space travelers were fired into space from a huge gun. Apart from the difficulty of entering an acceptable orbit in this way, there is the drawback that the travelers would first be flattened by the force of the explosion, and then fried as their capsule was heated by the atmosphere. The drawbacks in Verne's scenario could have been guessed even in his own lifetime. But, more generally, all attempts to look forward are influenced by what is known at the time of the forecast. Take one of the most famous science-fiction stories of all—*The War of the Worlds*—which appeared at the end of the nineteenth century. Its author, H.G. Wells, pictured a future in which Martians invaded the Earth. In his day— unlike today—this was not too great a leap of the imagination. Percival Lowell, a famous American astronomer of the time, claimed that observations of Mars showed its surface to be covered with canals. Their existence indicated not only the presence of intelligent life on Mars, but also that the inhabitants were running short of water. It was a small step from there to supposing that the Martians must be casting envious eyes on the large water resources that we have on Earth.

In fact, there was a wider assumption behind the belief in intelligent Martian life. As soon as it became apparent four centuries ago that the Moon and planets are similar sorts of bodies to the Earth, it was assumed that they must provide homes for living beings. The history of space exploration since then has been to narrow down the likelihood of life elsewhere in the solar system until now we are desperately hoping to find evidence of even the very simplest forms of life on Mars. One way science fiction has reacted to this disappointment is by peopling other planets, or their moons, with human colonists. Even so, science-fiction writers have usually proved too optimistic. Many stories written in the 1950s expected that colonies elsewhere in the solar system would have been founded by now. At a different level, the lack of life in the solar system can be countered by asserting that life, and especially intelligent life, may exist on planets circling other stars. The underlying belief here is that other planets in the solar system do not have life because they do not have suitable environments. Provide a suitable environment elsewhere and life will inevitably appear. It is further assumed that, so long as there are no major disasters, evolution will eventually produce on such planets lords of creation like ourselves. Indeed, many scientists, as well as science-fiction writers, believe that intelligent aliens would even show some similarity to our own basic body plan. The argument is that, for example, having two hands with fingers and thumb is about as good for manipulating objects as evolution could manage. So some system like it would be needed by any intelligent life form. Yet all these assumptions are open to question. For example, it is still far from clear what kind of conditions are necessary in order to produce life in the first place; nor is it clear how inevitable the course of evolution leading to us has been.

One maverick suggestion for getting round the bottlenecks in the production of life on a planet is to suppose that the space between the stars is filled with simple life forms, such as bacteria. These can drift down to the surface of any planet they encounter and, if the conditions are suitable, survive and propagate. The idea—which has been around for a considerable time—has been labeled *panspermia*. It has the great virtue of not requiring life to be generated afresh on each planet. Unfortunately, it also has major drawbacks. Stars are a long distance from each other. It would take bacteria millions of years to drift from one planetary system to another. During that time, they would be bombarded by all the radiation (such as x-rays) and the fast-moving particles that fill interstellar space. In laboratories on Earth even the toughest bacteria can be destroyed by such radiation, so there is a large question mark here over how long they could survive in space. Moreover, there would need to be an enormous number of them in circulation in order for at least one to reach the occasional planet capable of welcoming life.

Though methods of detecting planetary systems round other stars are rapidly improving, nearly all of the planets found so far are more like Jupiter than the Earth. Detecting Earth-like planets remains very difficult, and there is no guarantee that, when found, they will actually harbor life. An alternative approach, circumventing these difficulties, is to try and detect intelligent life elsewhere by looking for the communication signals that an advanced society might be expected to produce. Television programs started on Earth in the 1930s. They have been leaking out into space ever since, and have by now reached all the stars nearest to the Sun. Presuming that any of these have inhabited planets, we can imagine the aliens there waiting breathlessly for the next installment of a terrestrial soap opera. We can invert this, and ask whether we can detect anything like our own TV signals from civilizations elsewhere. The first attempt to use radio telescopes to detect such signals from other planets was made in 1960. The whole approach was put on a firm basis about twenty years ago with the establishment of the SETI (Search for Extra-Terrestrial Intelligence) project. Examining stars that are likely to have planetary systems, and analyzing the radio noises that come from their directions, consumes not only a lot of telescope time, but also a lot of computing power. Nowadays, anyone who wishes can become involved by letting the SETI project use some of their spare computer capacity for processing the data. To date, no recognizable signal has been detected. This does not necessarily mean that there is no-one near us. Astronomers have naturally been looking for radio signals of the sort that we ourselves use for communication. But would other civilizations, perhaps more advanced than ours, necessarily use radio for communication? We can only grope ahead, hoping that aliens do indeed communicate in much the same way that we do.

Going Places

Science-fiction writers get round the problem of communicating with aliens by dispatching their intrepid astronauts on space voyages to the distant stars. In the real world, this raises a problem that is as yet unsolved. Modern science accepts that the fastest anything can travel is the speed of light: in practice, the highest speed our current space probes can achieve is far below this limit. Outer space is huge. Consequently, once outside the solar system it takes a very long time for you to get anywhere. With the kind of technology currently feasible, it would take over 100,000 years to reach the nearest star. Science-fiction writers get round this in various ways. One is to accept this estimate, and to dispatch a huge spaceship which acts as a kind of substitute Earth, with generations of inhabitants living and dying before their destination is reached. Alternatively, the astronauts are somehow put into a state of suspended animation throughout the whole period until they near their goal—an idea that formed the basis for the book *2001* and the subsequent film. This has the advantage that it only needs a much smaller spaceship. But perhaps the most interesting approach is to suppose that a space drive is built—as, in principle, it could be—which gradually accelerates a spacecraft to very nearly the speed of light. One of the consequences of relativity theory is that time slows down for travelers who are approaching the speed of light. This leads to what is often called the "twin paradox." One twin leaves the Earth and is accelerated away into space. After a trip round our neighboring stars, he or she returns to Earth only to find that the twin left behind has aged much more. For the first twin, maybe two to three years have passed; for the twin remaining on Earth maybe twenty to thirty years. In fact, a space probe that gets close enough to the speed of light could traverse much of the known universe in the lifetime of a single astronaut. Unfortunately, when it returned to the solar system, many billion years of our time would have passed, and there would be no Earth left to land on.

Tales of interstellar travel that try to remain more or less within the boundaries of current science obviously need plenty of time to play with. Writers and readers of science fiction usually want it all and they want it now. In other words, they want their heroes and heroines to move rapidly from one part of the universe to another, and to come back to their starting point not much older than when they started out. In most stories, some device—such as "hyperdrive"—is introduced which allows the astronauts to evade the limitations imposed by the speed of light. Attempts to explain how such a device might work tend to be vague. One possibility that has received some scientific attention is whether or not "wormholes" might exist. According to relativity theory, space can be distorted by the presence of dense, massive bodies. If we imagine space as being like a large, flat elastic sheet, then putting a heavy body on it would depress the sheet at that point into a kind of pouch. This distorts space, as represented by the sheet, at one particular point. But the universe has many massive bodies, so space may be distorted over wide areas, as well as at particular points. Imagine now that this general distortion means that the elastic sheet is folded back on itself, a bit like the cover of this book. You can go from a point on the front cover to a point on the back cover by tracing a path from one to the other across the spine. But you can also drill a hole through the book from front to back, and this will give you a shorter route between the two points. Put these two pictures together. If the elastic sheet of space is bent back on itself, a pouch on the "front" may line up with another on the "back." The two can then merge, producing a hole in space connecting the two points, though they may be far distant places when measured in the usual way. We have created a wormhole. Jump down such a wormhole, and you will have the equivalent of hyperdrive: a way of moving very quickly between two points that are a long way from each other.

Unfortunately, there are drawbacks. Wormholes are likely to be very narrow. Moreover, they are very short-lived: immediately after the pouches join up, they part company again. Even if you could slip into a wormhole, you would immediately be squeezed out of existence. Cosmologists have explored the possibility of finding a material that could be used to line the wormholes and keep them open long enough for travel through them to occur. The properties required of such material have caused it to be labeled "exotic" (and when cosmologists call something exotic, they really mean it). There is as yet no prospect of finding such material. Outside science-fiction stories, we are still stuck for methods of rapid transit in the universe.

Science-fiction stories actually acted as one of the stimuli for the recent discussion of wormholes by cosmologists. But unfortunately science fiction tends to be rather disappointing from an astronomical viewpoint. It is great for telling us how humans (and aliens) might change and interact, but it typically tells us rather little about how the physical universe will look in the future. When science-fiction writers do attempt this, they face the usual problem that they have to rely on what is known at the time they are writing. Sometimes that knowledge has proved to be well-based. For example, in that early classic, *The Time Machine*, H.G. Wells is mainly concerned with the future evolution of human society, but he does describe the results of some astronomical changes, such as the slowing down of the Earth's rotation. The reality of this slowing down is still firmly accepted today. Another early classic, Olaf Stapledon's *Last and First Man*, is mainly concerned with the future evolution of humans and human society, but it also discusses astronomy. However, his timescales for future changes in the solar system were badly out. Stapledon was writing in 1930, when there was no good estimate for the lifetime of the Sun. The reason was that no-one then was sure how the Sun got the energy that kept it burning. In consequence, he overestimated how long the solar system would be habitable by human beings by a considerable factor. We can make this criticism firmly because we are fairly sure that we do know now how the Sun works. This is one of the areas of astronomy that has moved in the last half-century from the highly dubious to reasonable certainty. What we know about the future Sun, and how we know it, is laid out in the next chapter.

1. The Heat of the Sun

Why start a discussion of the future of the universe by looking at the Sun? Two reasons—one obvious, the other less so. The obvious one is that we and the rest of the solar system depend on the Sun to keep us going. If the Sun changes in the future, then everything in the solar system will be affected. So, before looking at what will happen to the various objects in the solar system, we have to predict what the future of the Sun will be. The less obvious reason is that the Sun is one of the basic building blocks of the universe. Wherever you look into space you see stars. All the assorted shapes and sizes of galaxies that the telescope shows us consist mainly of stars keeping company together. In fact, the universe can be pictured as a kind of town, built of bricks. In a town, the buildings vary in size and appearance; they are arranged in a network of streets and roads; but behind all the apparent variety, everything consists basically of bricks. In a similar way, stars, such as the Sun, are the basic astronomical building blocks, which provide much of the structure in the universe around us. As it happens, the Sun is a fairly average sort of star. This means that, if we can predict what will happen to the Sun in the future, we can also say what will happen to vast numbers of other stars. This, in turn, means that we can say something about the future of galaxies. So, starting with the Sun, we can look at the future of the universe in two ways: going down the size scale to smaller things, like the planets, and going up the scale to bigger things, like galaxies.

The Sun dominates the solar system. It actually contains 99.8 percent of all the material in our immediate locality. Planets, comets, and everything else only make up 0.2 percent. For astronomers, the most important characteristic of an object is usually how massive it is. The reason is that gravitational pull is a major factor in determining what happens in the universe. In terms of our ordinary everyday units, the mass of the Sun is some 2,000 billion billion billion kilograms. (A billion here means one thousand million.) In comparison, the Earth weighs in at a meager 0.000003 percent of the mass of the Sun. Size is another important factor. If we could look at the solar system from outside, it would be obvious that the Sun is also very large compared with everything else in its vicinity. Suppose some giant hand moved the Earth to the center of the Sun. Our Moon, which currently swings round the Earth at a distance of some 400,000 kilometers, would then lie well below the Sun's surface. These figures put us terrestrial inhabitants firmly in our place. So, accepting that the Sun's future comes first in terms of significance, how can we investigate the way it will change?

Where the Sun Gets Its Energy

The first question to ask is: why should we expect the Sun to change at all in the future? The answer lies in the enormous amount of energy of all sorts—heat, light, and so on—that the Sun is continually pumping out. The total comes to nearly a billion billion billion watts per second. To put it another way, the energy coming from a few square meters of the solar surface would provide for the lighting and heating needs of an entire town. All this energy must come from somewhere: it is a basic rule that you cannot get something for nothing. So, where does the Sun get its energy, and for how long will the supply last? Ordinary ways of getting energy—like burning wood in a fire—will not do. Despite the great size of the Sun, it would burn out totally within a few thousand years if it had to rely on this kind of source for its energy. There is plenty of evidence to show that the Sun has not changed much over a far longer period of time than this. After all, major civilizations with written records go back several thousand years, and their writings give no indication of significant changes in the Sun. We need an alternative to this kind of simple burning.

A more likely source for the Sun's energy is its own gravitation. A massive enough object exerts an appreciable gravitational pull at its surface. The Earth may be relatively small, but the gravitational pull at its surface is more than enough to prevent us jumping off into space. The much more massive Sun produces a much bigger gravitational pull inward. The Sun is basically a hot ball of gas. Gas can be compressed, so the gravitational pull inward on the Sun's surface should cause it to shrink in size. (This is where the Earth differs from the Sun. The Earth is solid, and can withstand the gravitational pull inward.) If a gas is compressed, it produces heat—just as pumping up a bicycle tire produces heat in the pump because air is being compressed in the pump cylinder. So, if the Sun is shrinking under its own gravitational pull, the resulting compression could explain where the heat is coming from. Calculations indicate that the rate of infall of the Sun needed to produce its current amount of energy is quite slow. It would only need to shrink by a few meters a year (compared with its present diameter of well over a million kilometers). A Sun that heats itself by shrinking could therefore easily last for several million years.

Unfortunately, this still will not do. Archaeological evidence does not help much over periods of millions of years, but geological evidence does. Studies of rocks at the Earth's surface show that the Sun has been giving out heat at something like its present rate for a very long time. Some of the oldest rocks are a few billion years old. Yet they seem to have formed in the presence of liquid water. This means that the heat received from the Sun at the time they were forming was already sufficient to keep the Earth's temperature at about the same level as it is today: neither much higher (or the water would become steam), nor much lower (or it would be ice). It follows that the energy output of the Sun cannot have changed too greatly over the past few billion years.

We can turn the contraction argument round. There is actually no evidence that the Sun is shrinking at all. Yet gravitation must be at work. If the Sun is not, in fact, changing its size with time, then there must be something in its interior holding it up. The obvious thing is an internal source of energy. Then, when gravitation tries to pull the solar material inward, heat from this source will try to push it out again. (In the same way, a hot-air balloon expands outward against the surrounding atmosphere when its energy source—the burner—is turned on, heating the air in the balloon.) If all goes well, the result could be a standoff, with the push outward just balancing the pull inward. This would explain why the Sun has apparently changed very little over long periods of time. The big question of course is: what is the source of heat inside the Sun?

The answer was worked out during the first half of the twentieth century. It had long been accepted that all matter consists of minute particles, labeled *atoms*. In the early years of the century, it was found that the atoms themselves have a structure, which is in some ways reminiscent of the structure of the solar system. In the solar system, the massive Sun is surrounded by a bevy of much lighter, circling planets. Similarly, each atom has a massive central *nucleus* round which orbit a number of much lighter *electrons*. The mass of the nucleus and the number of electrons vary from element to element. For example, the two lightest elements are hydrogen and helium. Their difference is that the helium nucleus is four times as massive as the hydrogen nucleus, and the atom has two electrons to hydrogen's one. Next, astronomers realized that, if the Sun's interior was sufficiently hot, many of the atoms near its center would be broken down. In a gas—unlike a solid—atoms can move around freely. As they move around, they hit each other. How fast they move depends on the temperature—the higher the temperature, the faster they move. At a high enough temperature, the force of the impacts becomes so great that the surrounding electrons are stripped away from their central nucleus. Instead, of remaining stuck together, nuclei and electrons can then wander about separately in the Sun's deep interior.

As nuclei speed around in the Sun's center, they are constantly hitting either electrons or other nuclei. Now comes the interesting part. If nuclei hit each other hard enough, they can fuse together to form a new kind of nucleus. For the lighter elements, this fusion produces energy. Is this the internal source of energy that we are looking for? The calculations proved to be difficult to make, which is why it took a long time to agree on an answer. But it now seems that the temperature and the density near the Sun's center are, indeed, sufficient for nuclear fusion to occur. The temperature at the center of the Sun is estimated to be well over 10 million degrees, and the material there is at least 150 times denser than water. As one of the astronomers involved in this work said to his critics: "If you find a hotter place, you can go to it." And, indeed, these conditions prove to be just what is needed to produce the actual amount of energy that we see coming out of the Sun's surface.

The major problem in fusing nuclei together is that all nuclei have positive electrical charges. Equally, all electrons have negative charges. Since unlike charges attract each other, electrons happily attach themselves to nuclei. But like charges repel each other. Electrons therefore repel electrons and, more importantly, nuclei repel nuclei. You need a high density and pressure to overcome this electrical repulsion. The hydrogen nucleus has the lowest electrical charge of any nucleus. It consists of a single electrically charged particle, usually called a *proton*. All other nuclei contain more than one proton, and, being more highly charged, therefore repel each other more strongly than hydrogen nuclei do. Hydrogen is also the commonest element in the Sun. As a result of their large number and low repulsion, hydrogen nuclei provide the source for the fusion currently going on in the Sun's interior. What happens is that four hydrogen nuclei interact with each other in stages. First two protons fuse together; then a third adds itself to them; finally a fourth joins in. In this process, two of the protons are converted into electrically neutral particles (*neutrons*) which act as a kind of glue to hold the protons together. A nucleus consisting of two protons and two neutrons is characteristic of helium. So the overall result is to convert hydrogen nuclei into helium nuclei. The conversion takes time, even under the conditions found in the center of the Sun, but the energy that comes from each transmutation is considerable. (This is why there have been continuing attempts to produce hydrogen fusion on Earth.)

Turning one element into another inside a star is often called "nuclear burning." For the Sun, the process is specifically "hydrogen burning." The conditions at the center of the Sun certainly mean that hydrogen burning must occur there: but can it keep going for the billions of years demanded by geologists? We can check this by estimating how much hydrogen there is in the center of the Sun, and how long it would take for it all to be used up, supposing that the Sun keeps burning at its present rate. The calculations show that hydrogen burning can, indeed, supply the energy needs of the Sun for many billions of years—in good agreement

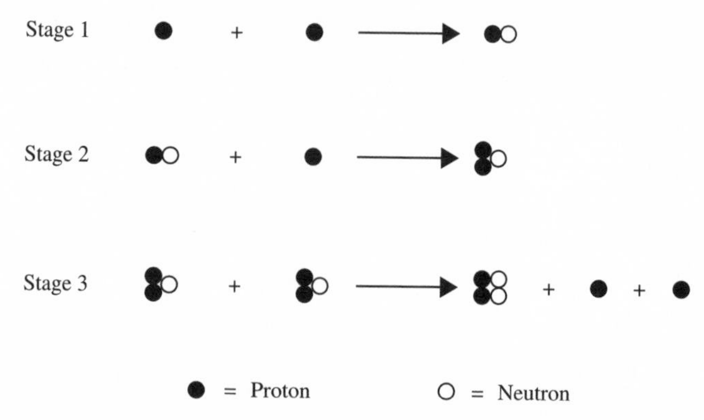

1.1 How hydrogen burns. Stage 1: Two protons (hydrogen nuclei) combine to form heavy hydrogen. One of the protons is converted into a neutron. Stage 2: The heavy hydrogen combines with another proton to form light helium. Stage 3: Two light helium nuclei combine to form normal helium, simultaneously putting two protons back into circulation. This is the simplest way of burning hydrogen to helium, but burning by other routes also occurs. Each of the steps in the process produces heat.

with the geological evidence. So we finally have our answer: the Sun gets its energy from nuclear reactions near its center.

How the Sun Reached Its Present State

The nuclear-burning answer explains satisfactorily why the Sun is stable now, but does not tell us what changes might occur. What of the distant past and the distant future? What happened before the Sun started burning hydrogen? What will happen in the future when the Sun has used up all its hydrogen supply? It might seem that only the second of these questions should concern us here. In fact, it turns out that understanding the past and present of the Sun and other stars is essential for a proper understanding of their future. This means that we have to look backward in time before we can look forward.

What observational evidence can we hope to find about the past and future of the Sun? One way is to examine other stars. They are known to have a spread of ages. So we can look both for stars that are younger than the Sun, and for others that are older. Comparing them may then give us clues to the life-history of the

Sun. This approach is like guessing how a tree develops by examining a forest. A walk through the forest shows trees at different stages of development—from young seedlings, through mature trees, to fallen tree trunks. From these different stages, we can put together a picture of how a single tree might develop. This is one way; there is another. Instead of an observational approach, we can try a different route, involving theory. In this, the Sun is considered to be a sphere of hot gas with a certain mass and composition. Computers can then be used to calculate how such a ball of gas will develop with time. In practice—as is usual in science—the two approaches go hand in hand, with the observations and the theory modifying and extending each other.

Theory and observation suggest the following answer to the first question— What was the Sun doing before it started burning hydrogen? The Sun started out as a big, diffuse cloud of gas and dust. The gravitational pull inward on the outer layers of this cloud made it contract, so that it became smaller and denser. This contraction generated heat, raising the temperature of the cloud. The outer layers of the cloud now acted like the insulation round a hot-water tank. They made it difficult for the heat generated nearer the center of the cloud to escape easily outward. Consequently, the center heated up more than the outer layers. The infall of material, and the consequent heating, continued until the center of the cloud became hot enough and dense enough for the hydrogen nuclei to start interacting. At this point, the heat generated by the hydrogen burning opposed any further contraction, and the former cloud became our own Sun. What is true of the Sun is also true of other stars: they started their careers as clouds of gas that contracted. Observations show that there are still some gas clouds in the Sun's vicinity. Indeed, one nearby cloud, in the constellation of Orion, is in the process of producing new stars at the moment.

How the Sun Has Changed

When the Sun was born, it obviously had the same composition as the cloud from which it was made. Once it started burning hydrogen, this composition changed: the helium content of the

Sun began to increase and its hydrogen content to decrease. The question is whether this alteration in composition has affected what we on Earth receive from the Sun. Have its brightness and surface temperature remained the same, or has the change in composition made a difference? Before this question can be answered, there are actually two other points that need to be got out of the way.

The first concerns which parts of the Sun are affected by the change in composition. Hydrogen burning occurs in the hottest region of the Sun—round its center. Does the new helium that is produced there mix in with the rest of the Sun, or does it stay where it is produced? As a look at photographs of the Sun's surface indicates, the outer layers of the Sun are boiling. The heat rising from the center produces currents, which bring hotter material from the Sun's interior to the surface. There it cools and falls back again to the interior. (In the same way, when a kettle boils, it is because the water at the bottom of the kettle is being heated. It rises, cools down again, and is replaced by new hot water from below.) This whole activity of transferring heat by boiling is called *convection*. When it occurs, it clearly mixes up all the material together. But theory suggests that this convection does not extend throughout the Sun. Only the outer 30 percent of the Sun is moving in this way. In the inner 70 percent of the Sun, the material is stationary. The heat from the center of the Sun still has to get out, of course, but now it passes through the solar material in the form of radiation. (Similarly, when you sit in front of a fire, the heat is reaching you as radiation, which passes through the air in the room.) In this central region of the Sun, there is no mixing of material. The hydrogen-burning zone is at the center of this stationary region. This means that changes in composition due to nuclear burning stay where they occur.

The second point concerns the present age of the Sun. For how long has the Sun been burning hydrogen at its center? The problem here is that we have no direct way of measuring the age of the Sun. This contrasts with the situation on Earth, where there are good methods for dating rocks using the radioactive materials they contain. Radioactive elements, which occur in small amounts in many rocks, are unstable. They break down into other elements at a regular rate (which is why they can be used as clocks). For example, uranium breaks down slowly, producing, amongst other things, helium. A measurement of the relative amounts of uranium and helium present in a rock gives some idea of its age. The more helium and the less uranium, the older the rock is. It seems safe to say that the Sun must certainly be older than any rocks on the Earth's surface, so radioactive dating on Earth gives us a minimum age for the Sun. Besides rocks from the Sun, we also have rocks from the Moon, together with meteorites (mainly fragments from the asteroid belt—see Chapter 5) which hit the Earth. Interestingly, the results from radioactive dating indicate that all these bodies were formed at the same time. It seems reasonable to guess that this was the time when the whole solar system, including the Sun, was born. If so, then the age of the Sun (and of the solar system) is about 4.5 billion years.

This information about internal mixing and age provides what is needed to calculate how the Sun might change with time. If the results of hydrogen burning have never reached the outer regions of the Sun, then the Sun's outer layers must still have the same composition as the cloud from which they were born. The composition of the Sun's surface today can be measured fairly easily. If we could dig out one kilogram of the surface layers, it would contain 710 grams of hydrogen, 270 grams of helium and 20 grams of everything else. If we turn these figures into numbers of atoms present, remembering that hydrogen is the lightest element, then the outer layers of the Sun consist of 92.1 percent hydrogen atoms, 7.8 percent helium atoms and 0.1 percent everything else. If the Sun as a whole started with this composition, then it obviously had plenty of hydrogen available for burning.

The interesting point comes if we calculate the properties of the Sun using this composition at its center. It becomes apparent that the predicted brightness and surface temperature do not agree with observations of the Sun as we see it today. The crunch question is this—if we allow for the fact that the Sun has been burning hydrogen for 4.5 billion years, will the calculated properties agree better with the observations? Answering that question is quite tricky. It involves constructing a series of theoretical models of the Sun to see how it has evolved with time. Even with powerful computers, this can take a lot of computer time. The first result the calculations give is the composition at the center of the Sun today.

A kilogram of material from there no longer contains 710 grams of hydrogen. Hydrogen burning over 4.5 billion years has decreased the amount to less than half: it now stands at 340 grams. The amount of helium has correspondingly increased.

How does this affect the properties of the Sun? From the point of view of a hydrogen nucleus at the center, meeting other hydrogen nuclei nowadays is becoming increasingly difficult, because there are fewer of them. Yet the Sun still has to hold itself up against the gravitational pull inward. So it responds by becoming hotter at the center. The higher temperature means that the nuclei move faster. This increases the rate at which they meet each other, and generates the necessary energy to keep the Sun stable. The resulting rise in the central temperature leads to increases in both the brightness and the size of the Sun. Calculations indicate that, over the lifetime of the Sun (from its creation till now), its brightness has increased by some 40 percent, and its size by more than 10 percent. Its surface temperature has also increased slightly.

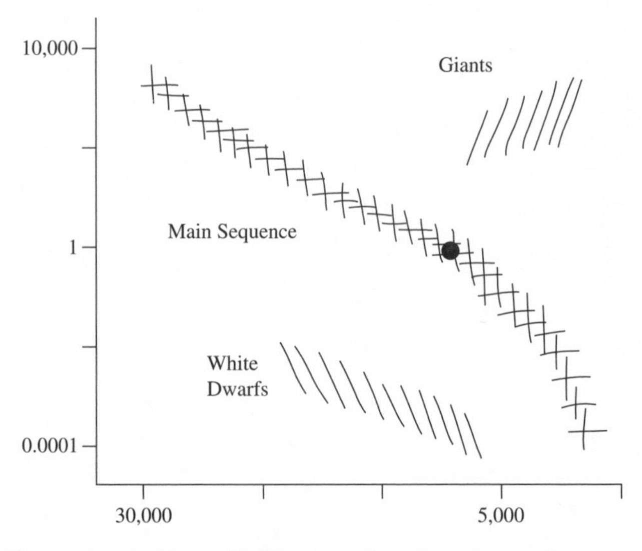

1.2 The Hertzsprung–Russell diagram showing the main groups of stars. The brightness of the stars increases as you move upwards in this diagram—from 1/10,000 of the Sun's brightness at the bottom to 10,000 times the Sun's brightness at the top. Surface temperatures (in degrees Centigrade) are shown running from low temperatures at the right to high temperatures at the left. This also corresponds to changes in the colors of the stars: from red at the right, through yellow, to blue-white at the left.

With these changes, the properties calculated for the Sun now fit well with the observations. In addition, we have an answer to our original question: is the Sun changing with time? The answer is yes, but very slowly.

The Sun's Next Stages

The easiest way of examining how the Sun evolves with time is to ask how its surface temperature and brightness are changing, as we did in the previous section. The reason is simply because these are the easiest properties to measure, not only for the Sun, but for other stars too. The surface temperature of a star determines its color, which can be measured quite readily. A hot star looks white-blue; a cool star looks red. The Sun lies between these extremes. It looks yellowish, corresponding to a surface temperature of some 5,500 °C. It is possible to arrange all the stars that are burning hydrogen into a color sequence ranging from blue at one end to red at the other. When you do that, you find that the hottest stars are also the brightest. So there is a link between brightness and color. Further investigation shows that the brightest, hottest stars are also the biggest and most massive. This makes sense. The more massive a star, the greater its gravitational pull inward. To resist this bigger pull, the star's interior must supply more heat. So the center of a massive star is hotter, and the hydrogen burning goes on faster than in the center of the Sun. Consequently, the heavier star is brighter, hotter and bigger than the Sun.

Astronomers often use a simple diagram to represent the properties of stars. It plots the brightness of stars in an upward direction against the surface temperatures in a sideways direction. The idea for this picture came from two astronomers in the early twentieth century—Ejnar Hertzsprung in Denmark and Henry Russell in the United States. It is therefore called the Hertzsprung–Russell diagram, often abbreviated to H–R diagram. Because brightness and surface temperature are related to size, the position of a star on the diagram also indicates how big it is. So the H–R diagram can provide a graphic illustration of how stars differ from each other in their properties. When all the stars that are burning hydrogen like the Sun are plotted on this diagram, they are found to lie on a band going from small, red and faint to big, blue and bright. This band is called the *main sequence*. The name comes from the fact that such stars are so common. The vast majority of the stars we can see in our neighborhood are, like the Sun, members of the main sequence. In fact, main-sequence stars are common throughout the universe, which is why the Sun provides an excellent model for understanding stars everywhere. As a result, if we can answer our basic question—how will the Sun evolve in the future?—what we find can be applied to very many other stars in the universe.

The theoretical calculations that have been made to determine how the Sun has changed over the past 4.5 billion years can be carried forward in time to find out what will happen to the Sun in the future. The results indicate that the Sun will continue to burn hydrogen for another 6 billion years or so. During this time, its brightness, size and surface temperature will increase at the same slow rate as at present. To put it another way, the Sun is less than half-way through its hydrogen-burning phase. Earth-dwellers have plenty of time left to worry about what will happen when the Sun runs out of hydrogen to burn. At the same time, by the end of this slow period of change, the Sun will be over twice as bright as it is at present. Somewhere along the line, this increasing brightness could create difficulties for our terrestrial environment. (Whether it will or not is discussed in Chapter 3.)

To understand what happens at the end of this long period of slow change, we have to look more closely at where exactly hydrogen is burning in the Sun. The rate at which nuclei burn is very sensitive to the temperature. Since the temperature and the density are highest at the center of the Sun, this is where most of the hydrogen burning occurs. But the regions round the center are also hot enough and dense enough to burn some hydrogen. As the interior of the Sun gradually heats up with time, so these regions round the center increase their contribution to the overall production of energy. By the time the final remnants of hydrogen at the center itself are burnt up, the surrounding regions are providing a significant fraction of the Sun's total heat. In the next stage, they produce still more. The Sun's center has been holding up the layers above it by producing heat from hydrogen. No hydrogen means no new heat, which means, in turn, that the center feels the full weight of the layers above it. They push the center inward, so that it begins to contract. But we know that contraction produces heat. In consequence, the regions round the center become hotter, and start to burn hydrogen much more quickly. Instead of burning hydrogen at its actual center, the Sun is now consuming it in a shell that surrounds the center.

The extra heat generated by the shell increases the Sun's brightness, and also makes it grow bigger. This leads to a big increase in the surface area of the Sun, which, in turn, leads to a fall in the surface temperature. So the Sun is now growing not only bigger, but also redder. As a result, it no longer fits satisfactorily along the main sequence. A main-sequence star that is red should be less bright than our present Sun, not brighter. When exactly the Sun will leave the main sequence is partly a matter of how the main sequence is defined. For example, we can define it in terms of the observations: when will the Sun clearly look different from all the other stars that we use to mark the main sequence? Whatever the definition chosen, within a billion years of having burnt up the hydrogen at its center, the Sun will clearly have ceased to be a main-sequence star. Its growth in size is now taking it toward membership of a quite different group of stars—the *giants*. Most stars fall into one of two groups—either dwarfs or giants—according to their size (with the dwarfs, not surprisingly, much smaller than the giants). Within these groups, there are various subgroups. For example, supergiants are stars that are much bigger than ordinary giants. The Sun, as we presently know it, is a main-sequence dwarf. Its evolution away from the main sequence will promote it to a quite different group of stars—the *red giants*.

During this stage in its evolution, the Sun has a core of helium, where no nuclear burning occurs. This is surrounded by a shell where hydrogen is still burning to helium. As time passes, the hydrogen-burning shell gets gradually thinner, but all the time it is producing more and more helium, which it dumps on the central core. Ultimately, such dumping leads to a problem. We know that a star is stable when it generates enough heat internally to balance the gravitational pull inward. This argument applies not only to the Sun as a whole, but also to its core regions. Though the helium core of the developing Sun is certainly hot, it is not actually producing any new energy. That is being done by the shell surrounding it. Sooner or later, as the helium core grows, its gravitational pull inward will exceed its ability to withstand the pressure from above. When that point is reached, the core collapses. As we have seen, contraction produces heat, so a rapid collapse produces a lot of heat quickly. This episode is called the "helium flash"—"helium" because that is the element involved in the collapse, and "flash" because the sudden heat generation causes the Sun to emit a lot of light in a short period of time. The new input of heat not only makes the Sun much brighter, it also makes it expand to many times its present size. According to current estimates, the flash will take it to some 170 times its present diameter, and 2,300 times its present brightness, before it falls back again. Suppose we imagine that our present Sun has the size of a grapefruit. Then, at this stage in its future life, it will have expanded to something like the size of a typical house.

The heat generated by the collapse has another result. It triggers off the next nuclear burning stage at the center of the Sun. We have seen that hydrogen burning requires four protons to be squeezed together to form a helium nucleus. The question is: when, in turn, the helium nuclei are pushed together, what new nucleus can they form? The obvious answer would seem to be a nucleus that consists of two helium nuclei. That is a nucleus with four protons and four neutrons. Unfortunately, it turns out that such a nucleus is unstable: it falls apart as quickly as you try to push it together. The only way of forming a stable nucleus is to fuse three helium nuclei together simultaneously. The result is a nucleus containing six protons and six neutrons, which is characteristic of that well-known element, carbon. (Sometimes the carbon will burn a little further, adding on another helium nucleus, to produce the equally well-known element, oxygen.) Collisions between two helium nuclei in the Sun are common. It is much less common for three to collide at once. For that to happen, the helium nuclei must be colliding very frequently indeed. That requires a very high central temperature—ten times as much as for hydrogen burning—along with a high density. So the Sun, in this part of its career, has a much denser, hotter core than before, but this is surrounded by very large and diffuse outer layers. Convection continues in the outer layers, as in the present Sun. Indeed, because of the expansion, convection will now extend through the outer three-quarters of the future Sun. But the important point is that the helium-burning region lies below this convective region. As a result, the continuing changes in composition at the center are still not mixed with the outer layers.

The increased input of heat due to helium burning at the center leads to an increase in the surface temperature, and the Sun's color now moves back from the red through the yellowish region of the spectrum. There follows a familiar story. The helium burns most rapidly at the center of the Sun. After a time, therefore, a core consisting of carbon and oxygen accumulates there. Correspondingly, helium burning moves to a shell surrounding this core. This diminishes the surface temperature and the Sun retreats back to the reddish end of the spectrum. Changes in the Sun are now occurring quite rapidly by astronomical standards. If

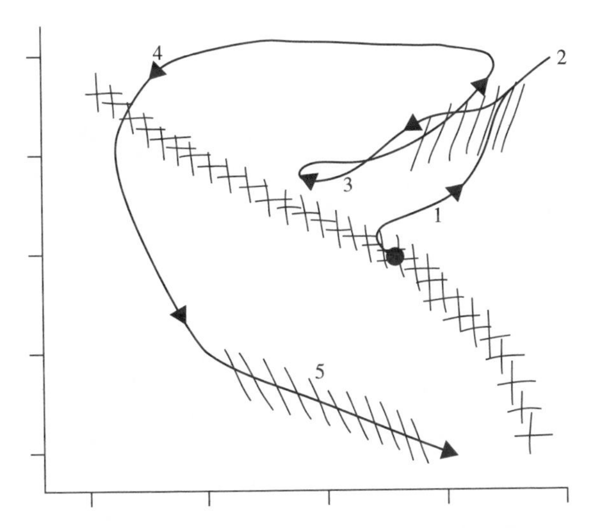

1.3 This H-R diagram shows how the Sun will change its appearance in the future as it evolves. The arrows indicate the direction in which it will move, but remember that it will pass through some parts of its path much more rapidly than through others. The main stages are labeled: 1 Hydrogen burning in a shell round the core; 2 Helium flash; 3 Helium burning in the core; 4 Planetary nebula; 5 white dwarf.

its changing position is plotted on a H–R diagram, it can be seen to be weaving its way backward and forward at much brighter levels than the main sequence, and with increasing speed. Its main-sequence lifetime lasts for some 10 billion years; its red-giant lifetime continues for hundreds of millions of years. But helium burning only provides sufficient heat for tens of millions of years. There are two reasons: burning helium produces less energy than burning hydrogen, and also, because of the high temperatures involved, the helium burning goes very rapidly.

The Sun's Final Stages

By this time, the material near the center of the Sun is very highly compressed indeed. The individual particles there find it increasingly difficult to move freely. In fact, the center becomes less like a gas, and more like a solid in its properties. Stellar material that acts in this way is called *degenerate* (which is not meant to be a moral judgment). Degenerate matter is better able to withstand pressure, and so less likely to collapse, than ordinary stellar matter. Consequently, the degenerate carbon/oxygen core that the Sun now develops is better able to stand up to the continuing pressure from above. However, degenerate material has two other properties. One is that it takes up a lot less space than ordinary stellar material. (In the same way, ice takes up much less space than steam.) The other is that nuclear burning can make it unstable. If nuclei start fusing together in degenerate material, it rapidly explodes back into ordinary material again.

The first point explains what happens next in the evolving Sun. When the core becomes degenerate, it decreases in size. Because the core shrinks, material further out in the Sun is drawn inward. As always, this fall inward heats up the new material. But these more distant layers of the Sun still contain hydrogen. Their rise in temperature means that this hydrogen starts burning in a new shell round the core. Now the second property of degenerate matter comes into play. The helium produced by the hydrogen-burning shell is deposited on the underlying degenerate core. But that core is still hot enough to burn helium to carbon and oxygen; so it will start burning the deposited material. In other words, the

Sun is providing combustible material to a zone of degenerate matter—an unstable situation. The result is an explosion: the size and brightness of the Sun both increase rapidly. It does not last. Once the new addition of helium has been burnt, the Sun settles back to something like its previous size and brightness. Hydrogen burning round the core resumes. But this leads to more helium being added to the core again, and so to another explosion. These explosions are again called "helium flashes." (Strictly speaking, they should be called "helium-shell flashes," to distinguish them from the original helium flash, which happened when helium started burning in the Sun's core.) A whole series of these flashes occurs, each one lasting for about ten thousand years over a total period of some hundred thousand years.

By the end of these explosive episodes, the Sun has effectively used up all the new energy that resulted from the formation of a degenerate core. Where else can it look for a source of heat? Carbon and oxygen can be burnt to provide energy, but, because their nuclei have larger electrical charges than hydrogen or helium, it is much more difficult to push them close together. This means that much higher temperatures and densities are required to burn them than are necessary for hydrogen or helium burning. So just how highly can the central regions of the Sun be heated and compressed? The answer depends on its mass. The more massive a star, the bigger its gravitational pull inward, and therefore the more its central regions can be compressed and heated. So how does the Sun measure up in these terms?

By this stage of its life, the Sun has actually lost a significant fraction of its original mass. When it first grew to become a red giant, its gravitational hold on its outer layers became less effective, allowing some of its surface material to stream off into space. In fact, the outer layers of the Sun at the red-giant stage become unstable. It can be predicted that they will pulsate in and out—like a balloon that is alternately blown up and deflated—with a period of 200 days or so. Here, theory and observation can be linked together. Stars of this sort—called *long-period variables*—are well known to astronomers. Observations indicate that most of them seem to be losing material to space, as the Sun will do. The Sun subsequently loses more of its outer layers as a result of the helium flashes. By the time that helium burning ceases, the Sun will have lost nearly half of the material that it has today.

With all these changes, the Sun now looks entirely different, not only from its present appearance, but also from how it looked as a red giant. Having lost much of its outer layers, the Sun is no longer red in color, but blue. In fact, the final major outbursts will probably turn the Sun into a *planetary nebula*. These are familiar astronomical objects, consisting of a dense, bluish central star surrounded by a large cloud of emitted gas. (Their rather odd name comes about because, through a small telescope, a planetary nebula looks rather like a planet such as Uranus.)

As a result of all this mass loss, the Sun's gravitational pull inward decreases. There is no way now that its central core can reach the temperature where carbon and oxygen will burn. With no new source of energy available, there is only one way out. As we have seen, degenerate matter can resist pressure inward (so long as it is not too large). So the inner regions of the Sun compact themselves until all but the outermost layers, where pressure is low, become degenerate. In this state it is stable despite the absence of a central heat source. Most of the original outer layers of the Sun with their hydrogen and helium have disappeared into space. The star that now remains consists mainly of a degenerate core of carbon. (There is a thin outer veil of hydrogen and helium left over from earlier days.) The Sun in this state can be identified as a member of another group of stars that observers know—the *white dwarfs*. Because degenerate material takes up much less space than ordinary stellar material, the Sun has now become very small. It is more like the Earth in size than the Sun we know today. To put it another way, if we represent its current size by a grapefruit, then, when it reaches the white-dwarf stage, it will be about the size of a typical flower seed. The transition to this end stage is rapid. From the final helium flash to the white dwarf stage takes only a few tens of thousands of years.

As the word "white" indicates, a newly born white dwarf still has a high surface temperature: it is whitish-blue in color. But, in the absence of any internal source of heat, the Sun has simply become a hot ember that cools off with time. Consequently, both its brightness and surface temperature fall. Initially, this occurs quite quickly in astronomical terms. At the start of its life as a white dwarf, the Sun is about one-tenth as bright as it is today. Within the following billion years, this falls to about one-thousandth of its original brightness. Subsequently, the cooling occurs more slowly. The Sun will take longer than all its previous lifetime before it finally becomes a black dwarf, with virtually all its heat and light gone. By that time, its degenerate carbon interior has become very much like a crystal in its structure. In this final stage, the Sun will be a black diamond (slightly polluted by other elements) in the sky. The American poet Robert Frost once began a poem with the words: "Some say the world will end in fire / Some say in ice." From the way its future has been described in this chapter, it seems that the second option is the correct forecast for the Sun. Yet, as a later chapter will show, given enough time, the first option may still have a chance of occurring.

2. The Mobile Earth

The bits of the Earth that interest us most are those we can see—the land, the sea and the air. But they are just hangers-on. The essential part of the Earth is what lies beneath our feet. This is what matters most in discussing the future of the Earth, and so this is where we have to start. As with the Sun, the first question must be: what is the Earth like today? The problem, of course, is trying to discover the properties of something that you cannot see into. The deepest boreholes that have been drilled still only penetrate an infinitesimal distance into the Earth. We need some kind of probe that can penetrate deep into the Earth's interior. There is such a probe, and it was identified a century ago: it is earthquakes.

Earthquakes and the Earth's Interior

Earthquakes are caused by rocks in the outer layers of the Earth moving against each other. When they do this, they shake all the surrounding rocks, producing violent vibrations. With large earthquakes, these vibrations spread throughout the interior of the Earth. In some ways, it is like dropping a stone into a pond. The stone produces ripples in the water, which spread out far away from the original point of impact. The speed at which such vibrations spread out from the site of the earthquake depends on the type of material through which they have to pass. For example, after the lunar landings, it became possible to study moonquakes. Like earthquakes, these are caused by movements in the interior—in this case, of the Moon. The measurements showed that the speed with which vibrations move through the surface layers of the Moon is very close to the speed with which vibrations move through green cheese in terrestrial laboratories. Sadly, this was because the lunar material has the same porous structure as green cheese, not because it is made of the same substances. In general, the speed with which quake vibrations move depends on both the structure and the composition of the rocks through which the vibrations pass. The Earth has a different internal structure and composition from the Moon, so the speed with which vibrations pass through the Earth differs from the speed on the Moon. But, in both cases, the speed can be related to the properties of the material through which the vibrations are passing.

Earthquake vibrations are usually called *seismic waves*. "Seismic" comes from the Greek word for an earthquake, so that is simple. But what about "waves"? Any regular vibration can be called a wave. The ripples that spread out when a stone is dropped into a pond provide a typical example. Their ups and downs are regularly spaced. The distance from the top of one wave to the top of the next—labeled the *wavelength*—is used to distinguish between different kinds of wave. But this is not the only possible difference: vibrations can be of more than one kind. The vibrations on the surface of a pond go up and down. But take hold of a spring, pull it out, and let it go. The spring will vibrate, but the vibrations will be along the spring, not from side to side. Both types of wave are common in the world around us. For example, light waves are of the first kind, while sound waves are of the second. Earthquakes produce many different vibrations. But they can be divided into those that go backward and forward, like a spring, and those that go up and down, like the ripples on a pond. Any vibration in the first group is called a P-wave: it is linked to **p**ush/**p**ull motions in the rocks around the earthquake. Vibrations in the second group are called S-waves: they are linked to an up-and-down **s**haking of the surrounding material.

Nowadays, instruments for detecting earthquake waves are distributed all round the Earth. Obviously, the further an instrument is away from an earthquake, the longer it takes for the seismic waves to arrive there. A comparison of the differences in times of arrival makes it possible to determine where the original earthquake occurred. But, in addition, the instruments show that P- and S-waves arrive at different times. The two kinds of wave move at different speeds through the Earth's interior because they are affected differently by the nature of the material through which they pass. Studying the waves can therefore give a clue about what the Earth is like inside. Comparing measurements from instruments at a range of sites leads to one particularly significant conclusion. P-waves and S-waves have an important difference in the way they react to their surroundings. P-waves can pass through any type of material, whether solid or liquid. S-waves can only pass through solids, not through liquids. As James Bond might have said: earthquakes prefer their liquids pushed, not shaken. This difference led to a major discovery about the Earth's interior quite early on in the study of earthquake waves. It was found that both the P-waves and the S-waves from an earthquake could be detected over much of the Earth's surface. But, in an area of the Earth on the far side of the center from the earthquake, no S-waves could be observed. To the scientific detectives of the day, the solution was obvious. The existence of this shadow zone must mean that the central regions of the Earth are liquid. A liquid sphere at the center of the Earth would allow the P-waves to pass through to the far side, but would blot out the S-waves. Measuring the extent of the shadow zone gives a direct estimate of the size of the liquid core.

Since those early days, the size and nature of the liquid core have been pinned down quite precisely. On average, the Earth's center is some 6,350 kilometers below the surface on which we stand. The top of the liquid core lies about 2,900 kilometers down. This means that the core extends nearly half of the way out from the center to the surface. But detailed observations of earthquakes have revealed a further refinement. Some waves that should be absent according to theory actually get through to the shadow zone. Only one cause for this seems possible. The liquid core must contain at its center a smaller sphere of solid material. The inner solid core acts on some of the earthquake waves passing through the Earth in such a way that they turn up unexpectedly in the shadow zone. This piece of detective work was carried out by Inge Lehmann in the 1930s. She was not only the first female to make a major contribution to the study of the Earth's interior; she was also a pioneer of such studies in Denmark. As she showed, the solid core extends out to about a fifth of the distance from the center of the Earth to its surface. That makes it about the same size as the Moon. So the structure of the Earth is quite complicated—a solid sphere at the center, surrounded by a liquid shell, which is, in turn, surrounded by a solid shell.

Alongside the question of whether the interior of the Earth is solid or liquid, there is another: what materials does the Earth consist of? Seismic studies help here too, since the speed of the waves depends partly on the composition of the material through which they pass. The solid inner core seems to consist mainly of iron, with less than 10 percent of other elements. Iron also predominates in the liquid core, but here it is mixed with twice as much of these other elements. The solid outer part of the Earth is divided into two parts—the *mantle* (the part of the Earth between the liquid core and the surface layers) and the *crust* (the thin surface layers on which we live). The mantle consists mainly of oxygen, silicon and magnesium, joined together in a variety of chemical combinations. The crust is much more mixed up chemically, but the overall result is that the crust is less dense than the mantle. Just as a cork floats on water, because it is less dense than water, so the continents and the oceans are floating on top of the mantle. Now we know what the Earth looks like inside, we can

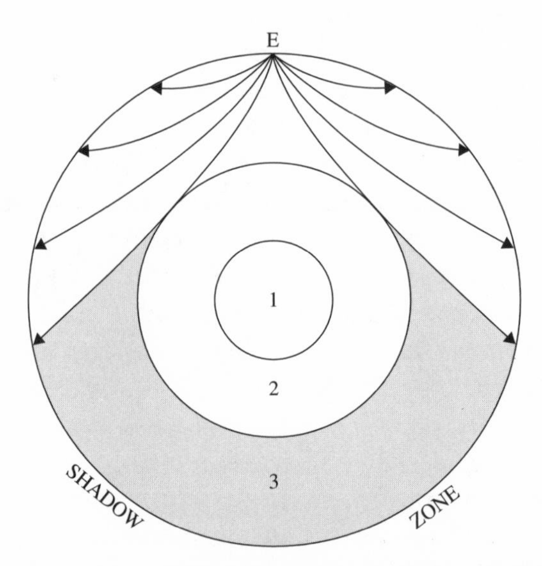

2.1 A cross-section of the Earth showing the passage of S-waves and the shadow zone. E is the site of the earthquake; 1 the solid inner core; 2 the liquid outer core; 3 the mantle.

turn to our main interest: is this structure changing with time, and if so how?

Changes in the Earth's Core

We started our discussion of how the Sun changes by asking where it got all its energy from. We can ask the same sort of question about the Earth. The Earth has a highly active surface—most spectacularly obvious when a volcano explodes. So what powers this activity? Volcanoes themselves, along with hot springs and the like, provide the clue. They all represent places where heat is coming up from below. Unfortunately, estimating how much heat reaches the Earth's surface from its deep interior is quite tricky. The reason is that there are other sources of heat near the surface that confuse the measurements. For example, radioactive materials, such as uranium, congregate just below the Earth's surface, and produce extra heat as they break down. After allowing for these various additions, the measurements show two things. Firstly, the center of the Earth must have a temperature of several thousand degrees. This is enough to make the core molten—confirming the deductions made from earthquake observations. Secondly, the Earth only loses heat quite slowly, because its outer layers form a good insulating blanket. So the Earth's interior, once it has become hot, takes a long time to cool down.

What do these results tell us about changes in the Earth's interior? Suppose we start at the center, and look first at what changes might be expected in the solid core. Whether or not the material near the center of the Earth is solid or liquid depends on the conditions there. At present, the pressure in the central parts keeps them solid despite the high temperature. Further out the pressure is smaller, but the temperature is still quite high. So the outer parts of the core are liquid. The heat measurements at the Earth's surface tell us that the Earth is cooling down. Unlike the Sun, the Earth does not have an internal source of heat: solids and liquids cannot contract and produce gravitational heating in the way that a gas can. So the temperature in the Earth's core is dropping. Since pressures remain the same, the region immediately round the solid core ultimately reaches a point where the temperature can no longer keep it liquid. What happens then is that the iron from the surrounding liquid core slowly condenses onto the central solid core. A residue of the lighter elements that are not carried down with the iron is left behind. This lighter material floats upward toward the mantle; its motion helps keep the liquid core continually stirred. The liquid core is, in fact, convective, with material moving up and down, just like the outer layers of the Sun.

The boundary between the outer core and the mantle seems reasonably stable, so that the core volume as a whole remains about the same. The growth of the inner core therefore means that the amount of space available for the outer core must be shrinking. Studies of the inner core suggest that it may have first appeared less than 2 billion years ago. In other words, the Earth existed for well over half of its lifetime up to the present without a solid core at all. Moving forward in time, the growth of the solid core will surely make it the dominant feature of the whole core in the future. But when? The rate of growth of this inner core is uncertain, but simple estimates suggest that it may take over much of the core by 3–4 billion years from now. The remaining liquid core will contain a much higher level of lighter elements than at present (the result of them being rejected by the solid core), but it will be less well stirred. Though—as we will see later—these changes in the core will produce some effects at the Earth's surface, they will have less impact there than parallel changes in the outer layers of the Earth.

Changes in the Mantle

The rate at which heat is lost from the central regions of the Earth depends on how good an insulator the mantle is. As with a hot-water boiler in a house, the better the insulation, the slower the heat loss. The rate of loss depends, in turn, on how the heat makes its way through the mantle. Since studies of seismic waves show the mantle to be solid, it might seem obvious that heat must pass through the material by conduction—the same way it does through the lagging round a hot-water boiler. But calculations show that heat is arriving at the Earth's surface faster than would be expected if it all comes by conduction. To match the observed rate, some internal heat must be transmitted through the mantle by motion—by convection currents like those in the Sun. So we have the surprising picture of hot bubbles rising through the Earth's mantle in the same way that the thermals beloved of glider pilots rise through the Earth's atmosphere. Yet the earthquake waves tell us that the mantle is solid. How can we square these two pieces of conflicting evidence?

The answer is a matter of timescale. There are substances which behave like solids on a short timescale, but flow like liquids on a long timescale. A good example is pitch, the black resinous substance used for many years to make wooden ships water-tight. If you hit a piece of cold pitch sharply, it will shatter. A blow is something that happens on a short timescale, and the pitch reacts as a solid. But in the longer term—say a century or more—the same block of pitch will flow slowly under the influence of gravity, changing its shape as it does so. The mantle behaves in the same way. Seismic waves pass through the mantle quickly, so it reacts to them like a solid. The input of heat to the mantle from the core takes place over billions of years. It is a slow process. The mantle responds to this long-term input by flowing, very slowly, like a liquid. "Very slowly" in this case means at the rate of about a centimeter a year.

The simplest picture of what is happening in the mantle supposes that currents rise from the edge of the core until they have nearly reached the crust; there they cool, and fall back again toward the core. But this is only one possible way in which convection might occur. Another way would be to have two layers in motion: one in the lower mantle and the other in the upper. On this picture, material heated by the core rises a certain distance into the mantle and then falls back again. Before doing so, it hands its heat on to the next region of the mantle, which, in turn, takes it up to the crust. If this sounds unnecessarily complicated, it should be added that some computer calculations actually suggest that both schemes may operate at different times, or even at the same time in different parts of the mantle. Besides this general scheme of convection, it seems that individual plumes of hot material can rise through the mantle in places, creating specific hotspots at the Earth's surface. Such events occur irregularly, but on average one might be expected every 30 million years—a short period in terms of the Earth's total lifetime. On occasion, these plumes can become very large—"superplumes"—and the molten rock they produce spreads over a significant fraction of the Earth's surface. One such superplume event seems to have started some 120 million years ago. Its effects at the surface gradually tapered off over the next 70–80 million years, but can still be detected by geologists under a considerable area of the Western Pacific.

Fortunately, the transfer of heat from core to crust seems to take about the same time—somewhere around a hundred million years—whichever way the material rises in the mantle. Moreover, how the heat affects the surface of the Earth is determined mainly by the nature of the mantle near the surface, which is the region we know most about. This simplifies forecasting what will happen in the future. The evidence from earthquakes suggests that there is a layer of rocks about 250 kilometers below the Earth's surface that can easily be distorted and made to move. The solid region above this layer is called the *lithosphere*. The part of the Earth we know best—the *crust*—floats on top of the lithosphere. It is thicker in continental regions (40 kilometers) and thinner under oceans (10 kilometers). The lithosphere is broken up into a number of fragments, called *plates*: a few big ones—thousands of kilometers across—and a larger number of smaller ones. Viewed from outside the Earth, it looks rather like a spherical jigsaw puzzle. Our continents and oceans are strewn across the tops of these plates. Now convection currents, whether in a boiling kettle or in the Earth's mantle, do not simply move up and down. After the hot material has given up its heat at the top, it is thrust aside by new hot material coming up from below. Consequently, the sequence of motions in convection is actually as follows: up to the surface; give up heat; move sideways; fall back down again. This means that the upper part of the mantle contains a lot of sideways motion. When it moves about, it drags the plates that lie on the surface with it. This reshuffling of the plates, and therefore of the continents and oceans on them, is known as *continental drift*.

The movements in the upper mantle push the plates about in different directions. Inevitably, they collide with each other. When this happens, one plate usually rides over the top of the other, pushing the latter down into the mantle. The Earth's gravitational pull encourages the descending plate to continue on its downward

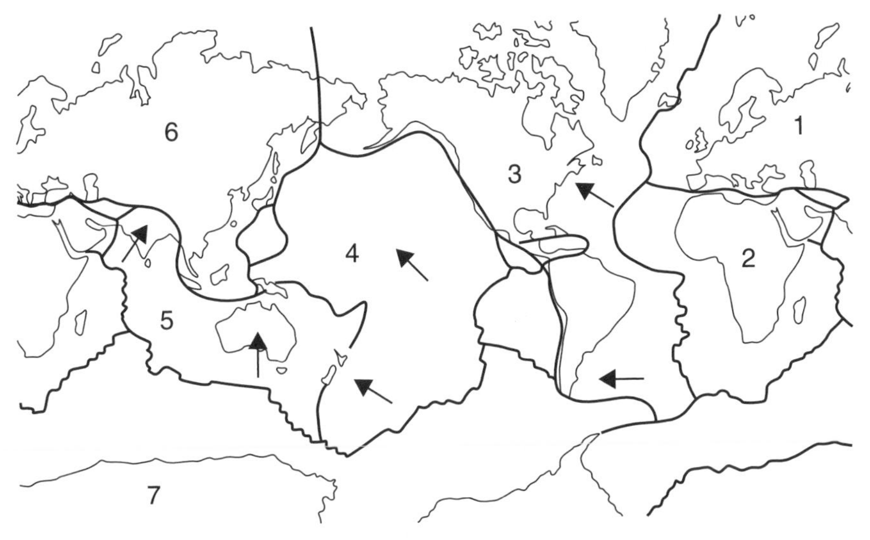

2.2 The main plates forming the Earth's surface (directions of movement are indicated by the arrows). The most prominent plates have been numbered: 1 Eurasian (western part); 2 African; 3 American; 4 Pacific; 5 Indo-Australian; 6 Eurasian (eastern part); 7 Antarctic.

path. As it slithers down, the other end of the plate moves away from the next plate along. The gap this makes between the two plates is filled by molten rock coming up from below. The new rock so created lifts the far end of the plate at the same time as the near end is being bent downward. The overall result is to make sure that the plate continues sliding downhill into the mantle. While the near end of the existing plate is therefore disappearing into oblivion, new plate is being formed at the far end. The overall result is that the surface layers of the Earth are always changing. The plates, as they are dragged about and interact, change their size, shape, and position. Since the continents and oceans are attached to the plates, it follows that they too must be constantly changing.

Continental Drift and Supercontinents

The existence of continental drift means that the Earth's surface must have looked different in the past and will again look different in the future. To try and say something about its future appearance, we need first of all to have some idea about how long it takes for major changes to occur. One clue to this is provided by the present rate of plate motions. Satellites now allow very precise measurements of how one plate is moving relative to another. Observing stations are set up on different plates, and their positions relative to a satellite are measured accurately at intervals. Over a period of several years, a comparison of these satellite distances gives an estimate of the speeds and directions of motion of the plates relative to each other. Average speeds are found to lie in the range 1–10 centimeters per year—a little more than the rate at which a human fingernail grows—and, as might be expected, similar to the rate at which material moves upward through the mantle.

To take one example, North America and Europe are attached to separate plates. The division between the two runs down the middle of the Atlantic Ocean. The two plates are currently moving apart. As they separate, new material comes up from below to fill the gap, so the Atlantic is constantly growing in size. Suppose that North America and Europe started out side by side and have moved apart continuously ever since. In other words, suppose that the whole of the Atlantic has been formed by repeated injections of new material from below. How long ago would the process have started? We know the present size of the Atlantic, and we can measure how fast the two plates are separating. So it is easy to estimate that the two continents should have started drifting apart some 200 million years ago. This estimate can be tested. If it is true that North America and Europe were originally close together, rocks taken from the ocean floor near the coasts of both North America and Europe should have formed about 200 million years ago, whereas rocks collected from near the middle of the Atlantic should be younger. That is indeed what measurements on the ages of the rocks indicate.

This method of looking backward can be applied to all continents and oceans. For periods up to 200 million years into the past, the geological record is generally quite good. When the same arguments are applied to the whole Earth, it appears that all our present continents, not only North America and Europe, nestled close together about 200 million years ago. One of the fits at that time—the one between the east coast of South America and the west coast of Africa—is so obvious that it has been remarked on for centuries. The resultant cluster of continents—a supercontinent—has been labeled *Pangaea* (meaning "all lands").

This surprising conclusion can be checked in another way. The Earth has been magnetic for billions of years, and its magnetism seems always to have run roughly in a north–south direction. When a molten rock solidifies in the presence of a magnet, it can capture some of the magnetism present. A rock cooling at the Earth's surface retains a magnetic imprint of the Earth's magnetism at the place where it formed. This, in turn, can give us some idea of how far north or south of the Equator the rock was at the time it solidified. It is as though the rock has been given its own magnetic compass, which always bears witness to where it was born relative to the Earth's magnetic poles. Measurements of the magnetism of rocks formed over the past 200 million years confirm that the continents were originally close together, but have since drifted apart. But they also give additional information. Pangaea, it appears, was an elongated supercontinent: it extended a long way both north and south of the equator. (In fact, North America, Europe, and Asia lay to the north, with most of the remainder to the south.) Magnetic measurements also help with the question of how long Pangaea existed. Their evidence suggests that the supercontinent only endured for a limited period of time. It accumulated when plates bearing continents came together, continued in existence for about a hundred million years, and then began to break up. We can now try to apply this picture of past movements to what will happen in the future. But before doing so we need to know more about the process by which continents come together and then break up again. First of all, how long have the individual continents existed?

As we have seen, continents float like corks above the turmoil that goes on below. As plates grind together beneath them, they have a good chance of surviving on top. So it is no surprise that radioactive dating shows that the continents have cores of ancient rock. On average, these cores have been in existence for half the age of the Earth. They are surrounded by a halo of younger, but still old rocks. In Canada, for example, the core regions date back 2.5 billion years, while the surrounding rocks are a billion years younger. The two lots of rock together are called the Canadian shield. There are similar shield areas in other continents. These shields are surrounded by belts of much younger rocks. So the central parts of continents have been about for a very long time. This is not true of the edges of continents. They have increased in size with time, as additional rocks have been added to them. It has been estimated that the total volume of continental crust has more than doubled in the last 2 billion years. Putting these two results together implies that the continents have existed for a long time, but they have changed their sizes, shapes, and orientations as time has passed. Since the changes happen relatively slowly, individual continents can be identified back to times before they came together to form Pangaea. For example, it seems that what is now North America—it looked different then—migrated a long way round the gradually forming supercontinent before finding its final resting place in the structure. When it split off from there later, it acquired more or less the size and shape we see now.

Detailed studies of rocks older than 300 million years, when Pangaea started to come together, are more difficult. However, they certainly show that continental drift has been operating for a long time. Indeed, there are hints that the whole process may be cyclical. The cycle starts with a supercontinent, like Pangaea. Over a period of 120 million years or so, this breaks up. The pieces wander away, creating new oceans (like the Atlantic) in the process. They reach a maximum distance apart after another 160 million years, then they start to move toward each other again. As a result, a new supercontinent is born after a further 160 million years. The whole cycle, from the breakup of one supercontinent to the breakup of its successor, takes about 440 million years on this picture.

Before we can apply all this to the future of the Earth, we need to be clear about one final question: why should a supercontinent, once it has formed, break up again? As with the plate motions themselves, the answer depends on the heat coming up from the Earth's interior. Heat flows more slowly through continents than through ocean floors because the continents are considerably thicker. So long as the continents do not cover too big an area, the heat held back under them can escape sideways round their edges. But a supercontinent acts as a heat blanket over a much larger area. Escape of heat sideways becomes more difficult, and much of it remains dammed up underneath. This accumulation of heat gradually stretches the continental crust above until the supercontinent finally breaks up into pieces. This is why the whole cycle starts again. The use of the word "cycle" does not necessarily mean that the activities recur regularly every 440 million years. As yet, the geological evidence for a regularly repeated cycle of continental drift is quite limited. But there is good backing for the belief that the continents came together and split apart again more than once before Pangaea appeared. Indeed, evidence for supercontinental splitting goes back for more than a third of the age of the Earth. The mechanisms involved in building up and breaking down supercontinents are straightforward, and can be expected to repeat. So it is reasonable to use the ideas we have been discussing here as a basis for looking into the future.

The Future of the Mobile Earth

We now know enough about continental drift to try and predict what will happen next. The predictions depend on the timescale that we choose. For example, some changes in the way continents look can occur over periods of a few tens of millions of years. The San Andreas fault in California is famous for its earthquakes. The fault actually represents the boundary between one plate border and another: the earthquakes are caused by the plates rubbing against each other. A long coastal sliver of California (including San Francisco and Los Angeles) lies on one of the plates, while the rest of the state is on the other plate. The coastal sliver is currently moving northward relative to the rest of California along the line of the San Andreas fault. A few tens of millions of years in the future, the sliver will form an elongated island separated from, and to the north of, present-day California. It will look a bit like a northern-hemisphere New Zealand (remembering that New Zealand itself was originally a single island). Europe is also in motion. For example, over the next 50 million years and beyond, the British Isles will be heading off toward Siberia and the north pole.

Instead of sliding past each other, plates can also hit head on. One example is the collision between the plate bearing India and the plate on which South Asia sits. The Indian plate is moving northward, and this is pushing it into the south side of the Asian plate. The collision began perhaps 40 million years ago. Since those early days, India has advanced some 2,000 kilometers to the north, pushing South Asia in front of it, and piling up the Himalayas in the process. India continues to move northward at a rate of some 5 centimeters a year, so the future here is predictable. The Himalayas will remain a major feature of the Earth's surface over the next few tens of million years, while India continues to smear itself out against Asia. In other places, the plate movements will produce mountains where they do not at present exist. For example, the Mediterranean Sea is being squashed between North Africa and southern Europe. In 50 million years' time, the present sea will be replaced by a range of mountains stretching the whole length of the Mediterranean.

Apart from these shorter-term changes (as geologists measure time), there is the big question of when we should expect the buildup of the next supercontinent. An examination of plate motions as they are happening now shows that most oceans are growing larger. The main exception is the Pacific Ocean. The plate bearing it is growing smaller as regions round its edge dip down, and are consumed under the neighboring plates (which bear the continents surrounding the Pacific). It is estimated that the North Pacific, for example, has shrunk by 13,000 kilometers over the past 150 million years. The obvious deduction from this would appear to be that the continental plates will ultimately form a new supercontinent in the area that is now the Pacific, since they are encroaching on it from all directions. However, there is evidence that things may be more complicated than this simple picture supposes. It may well be that the Pacific will open up again as the encroaching plates stop, and then retreat. If this happens—the Pacific widens while the Atlantic closes—a new range of mountains will appear on the east coast of America, while the existing mountains on the west coast will be eroded away.

The continents are currently approaching their maximum distances apart on the Earth's surface. This means that whether they advance further into the Pacific, or retreat back, their future motions must bring them closer together again. So whichever picture is correct, the present motions of the plates suggest that they should be nestling together once more in about another 200 million years. This fits in well with the idea that supercontinents form, break up, and reform in a cyclical way. But where will the new supercontinent lie on the Earth's surface? Looking at the present motions of the various plates, it becomes apparent that Africa is the one big plate that has moved rather little since Pangaea broke up. The next supercontinent may therefore accumulate around the African plate. This does not mean that the jigsaw of plates will fit back in exactly the same way as last time. For example, the bulges of South America and Africa will not slot back together again. Next time round, South America is likely to be further south, squashed up against the southern tip of Africa, and it will be North America that fits in higher up.

The basic driving force behind all these activities is the heat coming up from the inner parts of the Earth. The Earth, like the other planets, is thought to be an accumulation of material left over when the Sun was forming. "Accumulation" means that the material fell inward and compacted itself into a ball. As we know from our look at the Sun, the ball must therefore have warmed up, as gravitational energy from the infall of material was turned into

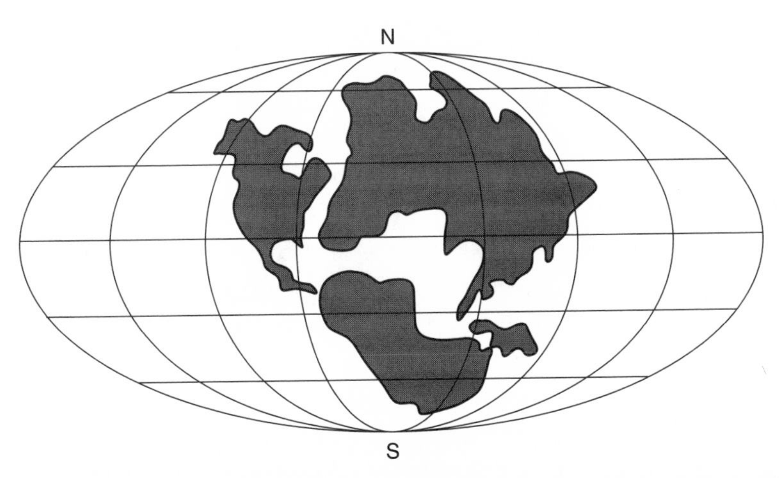

2.3 The next supercontinent coming together some 200 million years from now. North America can still be distinguished, but South America, Antarctica, and Australia have amalgamated in the southern hemisphere. Similarly, Africa, Europe, and Asia have united to the north.

heat. The residue of this original heat is still the main heating source in the Earth's interior, but its amount is gradually decreasing with time as it slowly leaks away into space from the Earth's surface. Since this heat drives the motions in the mantle, convection there will become slower and less powerful as time passes. This means that the movements of the plates will slow down, so that both the formation and the destruction of supercontinents will take longer. In addition, the pattern of convection in the mantle may change if the heat input becomes less. If this happens, it may become difficult to push all the pieces together into a supercontinent, and then to split them up again. The final cooling down is unlikely to go smoothly. As we have seen, continental areas are continually adding material. The crust will become thicker, making it harder for the heat to escape. When it finally does so, the consequences at the surface may be more violent. There will be major hiccups in the mantle—and so at the surface—as the Earth tries to get rid of its remaining heat. More "superplume" episodes are likely to occur, leading to repeated large outflows of lava at the surface.

Ultimately, convection will cease, and the remaining heat in the interior of the Earth will be lost by conduction outward. Clearly, plate motions will cease when convection in the mantle stops. But there is another factor at work which may halt continental drift before then. Plates slide down into the mantle relatively easily because the water present in the seas and oceans acts as a lubricant. As we will see in the next chapter, this water may disappear within about a billion years from now. If so, the whole mechanism will rapidly clog up because of the increased friction. Continental drift as it presently occurs must then stop, even though some convection will still be found in the mantle. If this estimate is right, then there will be time for perhaps three more supercontinental cycles before things grind to a halt. Because the continents will continue to grow, each supercontinent will be bigger than the one before.

While all this is going on in the mantle, the solid core at the center of the Earth continues to grow. Eventually, even the remnants of lighter elements left around it must also solidify. The end state of the Earth will therefore be as a cold, entirely solid body. The consequences of these cooling-down processes should be obvious at the surface by 3 billion years from now, but it may be several times longer than that before the Earth reaches its final stage of stagnation. At this point, both P- and S-waves will be able to travel side by side throughout the whole Earth—but there will be no earthquakes to produce them (and no-one to make the measurements).

3. The Earth's Oceans and Atmosphere

Movements in the Atmosphere

When we think of the atmosphere, we think of winds. But winds are just a local reflection of the global circulation of the atmosphere. The amount of heat coming up from the Earth's interior to its surface is very small compared with the amount of heat that the surface receives from the Sun. In consequence, it is solar heat that drives all the motions we see in the atmosphere and the oceans. Suppose that the Sun is overhead at the equator. The surface of the Earth must be heated more there than elsewhere. This means that the air above it is also heated more. As a result of the heating, the air becomes less dense and rises (like a hot-air balloon). It continues upward for a while until it reaches a stable layer of the atmosphere, called the *stratosphere*, at a height of about ten kilometers. This forms a barrier that prevents further motion upward. As a result, the air north of the equator is deflected sideways and starts moving northward. (Movements in the atmosphere south of the equator are mirror images of movements in the northern atmosphere. So heated air south of the equator correspondingly moves southward.) As this warm air moves to higher latitudes, it loses its heat and becomes denser again, making it fall back toward the Earth's surface. It has one more journey to make. The next lot of air at the equator has by now been heated and risen upward. The old air must therefore flow back to the equator to fill the gap. This is the same kind of round-and-round flow that we have seen before, and called convection. The result is a continuing circulation that produces northerly winds in the upper atmosphere north of the equator, together with southerly winds at the surface. An atmospheric motion of this kind is called a *cell*.

A similar argument can be applied to the terrestrial poles. The poles have low temperatures because they do not get much sunlight. Consequently, the air above them becomes cold and dense. It sinks down to the Earth's surface, pushing away the air already there. This produces winds blowing southward at the north pole (and conversely at the south pole). The polar air flows down to lower latitudes, where it warms up, rises, and then moves back toward the poles. So a global circulation is also created round each of the poles. It follows that air in the northern hemisphere, both at the equator and at the pole, should flow northward in the upper atmosphere and southward at the surface. But that is not the entire story. To see why requires a diversion.

If you stand at the equator and fire a gun northward, the shell appears to veer off to the right (eastward). What is happening is that the speed with which the Earth's rotation carries you eastward is different at different latitudes. At the equator it is maximum; at the poles it is zero. In between, it obviously decreases as the latitude increases. So a shell starting from the equator moves eastward with the same speed as the equator. At higher latitudes, the Earth's surface is moving less fast eastward than the shell. Consequently, the shell appears to veer eastward relative to the surface. This tendency for objects moving north–south to acquire also an apparent east–west motion is called the *Coriolis effect*. (Gaspard Coriolis was a nineteenth-century French mathematician who made a detailed investigation of the effect.) It obviously affects north–south currents in the atmosphere, or in the oceans, just as much as shells from guns.

We now have to revise our picture of how the atmosphere moves about. Suppose we think of air that is moving southward across the surface of the northern hemisphere. The Coriolis effect will force the winds that are trying to go southward into actually moving toward the southwest, because they are venturing into areas where the east–west speed of the surface is higher. Similarly, any winds that try to blow toward the north will actually move toward the northeast. The overall result is to complicate the way in which the air circulates. The Earth's atmosphere breaks up into three cells in each hemisphere. Along with the one near the equator and the other near the poles, there is a third (acting as a kind of roller bearing between them) at middle latitudes. In terms of surface winds in the northern hemisphere, these cells give northeasterly winds (popularly known as the trade winds) from 0° to 30°, southwesterly winds from 30° to 60°, and northeasterly again from 60° to 90°. (Remember that mariners always refer to winds in terms of the point of the compass from which they blow, not in terms of the direction toward which they are moving.) This is the simple picture. The reality is more complex. The picture of winds all blowing parallel to each other proves unstable. Instead, they break down into atmospheric whirls—cyclones or anticyclones, depending on which way they are blowing.

There are also seasonal effects. The Earth's equator is inclined at an angle of some 23.4° to the orbit of the Earth round the Sun. The result is that the Sun is only truly overhead at the equator on two days a year (the *equinoxes*). In between times, the Sun is overhead at different latitudes north and south of the equator. Its heating effect consequently varies throughout the year, producing summer and winter alternately in each hemisphere. Sorting out how the atmosphere moves nowadays is hard enough: predicting how it may change in the future is worse. Before trying to do so, we will compare what happens in the oceans.

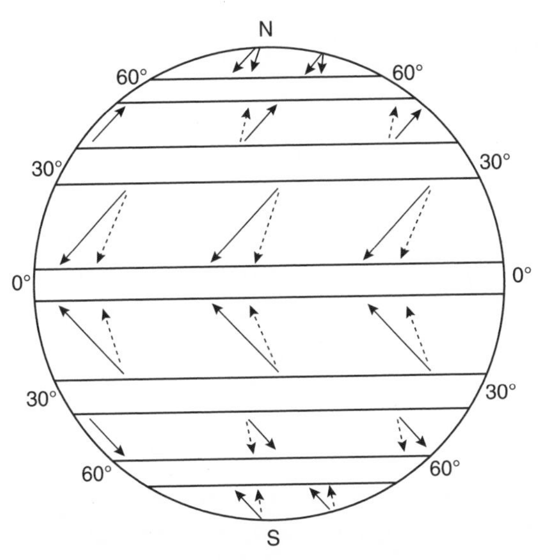

3.1 The basic circulation of the Earth's atmosphere, showing the surface winds. The dotted arrows indicate how the winds will change as the Earth's spin diminishes.

Movements in the Oceans

Ocean currents are influenced by the same factors as the winds. But for them the picture is even more complicated. The problem is that there are two ways in which water circulates in the oceans, and they only link up occasionally. The reason relates again to the source of heat—the Sun. Both the atmosphere and the oceans are heated by the Sun, but they are heated in different ways. Since the atmosphere is fairly transparent to sunlight, most of the light that is not blocked by clouds gets down to the Earth's surface. There it is absorbed by the surface, which heats up. The surface then pumps this heat into the atmosphere. The oceans, on the contrary, are fairly murky, so that sunlight only penetrates for a short distance. It gets nowhere near the ocean floor. Now convection depends on the heat coming from below. Put the heating element in an electric kettle at the top of the kettle, instead of at the bottom, and the water will not boil. Because the atmosphere is heated from below, it can convect and mix itself up thoroughly. For the oceans only the surface layers are strongly affected by the Sun's heat. In fact, the currents, produced by the Sun, are typically confined to the top 200 meters or so. The deeper layers below are left to go their own way.

The surface layers of the ocean are not only heated by the Sun; they are also stirred by the winds. Indeed, they are dragged along by the prevailing winds. But, unlike the winds, ocean currents are deflected by any coastlines they encounter. This leads to a more complicated pattern of circulation in the surface layers of the oceans than the basic pattern in the atmosphere. A quite different process is at work in the deep oceans, though currents there are equally affected by the distribution of the continents. The driving force is the production of cold salty water at the poles, which sinks to the bottom, leading to currents moving northward from Antarctica and southward from the Arctic. These deep currents only affect the surface layers of the ocean at a limited number of points. Consequently, it is the surface of the ocean that has most influence both on the atmosphere and on terrestrial life.

The way in which the ocean currents move will obviously be affected by changes in the distribution of the continents. When the continents are clustered together into a supercontinent, the remaining area—some two-thirds of the globe—forms a single gigantic ocean. The movements of the surface currents simplify under these conditions. A long stream of water flows along the equator, splitting into large loops that move north and south as the stream encounters the supercontinent. The pattern becomes more complex as the supercontinent breaks up. It is partly a question of whether enough space exists in the newly forming oceans to allow large loops of current to form. For example, the Gulf Stream, as we know it at present, is a large loop of warm water moving up the east coast of North America, across the Atlantic, and down the west coast of Europe. It has an important influence on the climate of these places. Yet it probably only appeared some 100 million years ago; prior to that, the Atlantic Ocean was not wide enough to allow such a loop to form.

How the currents change after a supercontinent breaks up depends on the exact way in which the jigsaw of continental fragments separates. When Pangaea split up, the northern and southern sections moved apart, leaving a clear seaway around the equator. This meant that large-scale currents could circulate freely round the equator without being diverted. The long time spent by the water in this region meant it became very warm. In turn, the oceans near the equator became warmer than they are today. This meant that they had more heat available to send off toward the poles, so warming the oceans as a whole. The situation remained the same for some 100 million years after the initial breakup of Pangaea. Then, as continental drift continued, North America joined with South America and Africa joined with Europe. The new layout not only effectively blocked the equatorial current, but also—by about 50 million years ago—began to open up a seaway in the south all round Antarctica. The resulting ocean current round the south pole was cold. Because currents no longer moved all the way round the equator, the heating coming from that region was weaker. It could not warm the new circumpolar current by very much. The increased cold near the south pole led to the Antarctic ice cap building up to the size we know today. Similar periods of warmer and colder currents will occur whenever a supercontinent breaks up. This in turn affects the atmosphere above the ocean, so producing changes in the climate.

Continental drift affects more than the currents in the ocean. It can also change the sea level. When a supercontinent splits, material comes up from the interior of the Earth to fill the gaps. The lines along which the plates split are marked by ridges where the hot material is rising. One example, as we have seen, is a ridge in the mid-Atlantic that provides new material from below to fill the gap as Europe and North America move apart. As the supercontinent breaks up into a number of separate plates, these ridges grow. They push away the ocean water around them, leading to a rise in the sea level worldwide of perhaps a few hundred meters. This may not sound much, but continents are typically surrounded by shallow margins. Because of this, a few hundred meters' rise or fall in the oceans can make a considerable difference to the amount of dry land. The movement of continental plates can even affect the nature of water in the interior of continents. When plate collisions are frequent, the impact of one plate on other leads to the creation of mountains. Water on mountain chains is typically found as ice or snow. When impacts are infrequent, low-lying land surfaces become common, and the water remains mainly liquid, forming lakes and pools.

Absorbing the Sunlight

Absorbing the energy arriving from the Sun is an essential first step in making use of it. A significant fraction of the light reaching us from the Sun is actually reflected directly back into space before it can be absorbed and used to warm the Earth. Clouds are good reflectors, but all surfaces reflect to some extent. (Otherwise, we would not be able to see them.) How much is reflected depends on the type of surface. For example, the amount reflected by the oceans is different from the amount reflected by land. Both reflect differently from ice or snow. As the continental fragments wander about the Earth's surface, the nature of the land, and of the water lying on and around it, changes. In consequence, the fraction of the incoming sunlight reflected or absorbed varies by small, but significant amounts. This leads to changes in the amount of heat absorbed by the Earth's atmosphere. So continental drift can have a direct effect on terrestrial climate.

Clouds are good at stopping light from the Sun reaching the Earth's surface. Currently, they reflect something like 20 percent of the incoming sunlight directly back into space. Where clouds are to be found depends, in part, on the type of surface down below. For example, one common cause of cloud formation over land is differences in height. When a wind hits the side of a hill or mountain, it is deflected upward. Any water vapor present is cooled as the air rises, condensing out as water droplets and so forming clouds. Mountain building is an essential part of the supercontinental cycle. When a supercontinent forms, the interaction between plates is especially strong round the edges where the supercontinent encounters the ocean. New mountain ranges are built up there, ringing the center of the supercontinent. These hinder the moderating influence of the oceans from extending into the interior. Consequently, the weather in the central parts of a supercontinent is more extreme than that found in separated continents.

As the supercontinental cycle pursues its course, there will be long-term changes in the global appearance of clouds as mountains form and then weather away. For example, the uplift of the Himalayas seems to have been a key factor in establishing the Indian monsoon. Clouds also form over the oceans. One example is the cyclonic clouds produced in tropical regions. These obtain their energy from the warm moist air that results from solar heating of the ocean surface. Such cyclones can extend over hundreds of kilometers; their typical whirlpool appearance is very familiar from satellite pictures. In the long term, tropical cyclones too depend on the layout of oceans and continents across the equator. When there is a clear seaway round much of the equator, the ocean currents become warmer. Cyclones should then be more frequent and longer-lived. People who suffer now from the effects of tropical storms may find some consolation in the thought that they could be worse.

Looking back in the geological record, the Earth's surface has clearly gone through periods of warmer and colder climates. The more extreme cold periods are usually referred to as ice ages (for obvious reasons). The timescale on which major ice ages have occurred actually fits in reasonably well with the timescale for continental drift. The first ice age for which there is good evidence occurred 800–600 million years ago, and was so extensive that it must have affected much of the Earth. Another, less extensive ice age occurred 460–430 million years ago, while the most recent one started 40 million years ago. But ice ages evidently depend on other factors besides continental drift. For example, an analysis of the most recent ice age shows that it intensified 2–3 million years back, and since then the extent of the glaciation has fluctuated up and down over periods of tens of thousands of years. We are currently enjoying one of the warmer times (called interglacials), which started some 10,000 years ago. The question we would obviously like to ask is—when will the next glacial period start? This cannot easily be answered by looking at past interglacials because they vary considerably in length. We need to understand why interglacials occur.

The most popular explanation was examined in detail in the early twentieth century by the Serbian mathematician Milutin Milankovitch. He pointed out that the spin axis of the Earth and the Earth's orbit round the Sun vary in a rather complicated way over periods of tens of thousands of years. This leads to changes in the way the Sun's heat affects the Earth. For example, one of the Milankovitch factors relates to the tilt of the Earth's equator relative to its orbit round the Sun. We have seen that this tilt is the cause of seasonal variations in the climate. Calculations indicate that gravitational interaction between the Earth and the other bodies in the solar system makes the tilt vary by a small amount—from 22° to 24.5°—over a period of around 40,000 years. Consequently, there are small variations in the way solar heat is distributed across the Earth, and this leads to minor changes in the seasons. In addition, the Earth moves round the Sun in a slightly elongated orbit. At those times of the year when it is closer to the Sun, it obviously receives more heat. One final factor is the orientation of the Earth's axis in space. This changes with time (which means that some thousands of years in the future, our present Pole Star will no longer mark the pole). Summing these, it is found that the part of the Earth which receives most heat varies with time, again on a scale of tens of thousands of years. When all the Milankovitch factors are put together, the result predicts fluctuations in the solar heat distribution with time which mirror moderately well the pattern of interglacials found in the geological record.

But some discrepancies remain. In consequence, some scientists doubt whether the Milankovitch factors can explain everything. Their suspicion is reinforced by the feeling that the changes in heating predicted by these factors are too small to explain the large changes in temperature from glacial to interglacial. Not surprisingly, such doubts have led to a number of competing theories appearing. However, new evidence suggests that the Milankovitch variations may be reinforced by other factors. Studies of the ice obtained from bore holes in Antarctica indicate that the amount of carbon dioxide in the atmosphere changes from glacial to interglacial periods. The amount of carbon dioxide—as will be explored below—can influence the surface temperature of the Earth. So variations in the amount could amplify the effects of the Milankovitch factors, and remove some of the discrepancies. If we accept the theory, then it is possible to estimate how much longer our benign interglacial will last. It seems that the Earth should experience a slow decline in temperature over the next few thousand years; but a full-blown ice age should be some tens of thousands of years in the future. The exact way in which the Milankovitch cycles work depends on the distribution of land

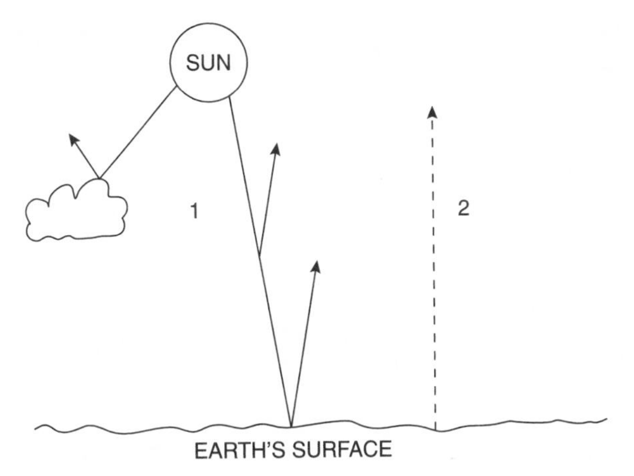

3.2 The greenhouse effect. On the left (1) is the incoming light from the Sun—20 percent is reflected back into space by clouds, and 10 percent by the rest of the atmosphere and the Earth's surface. The remaining 70 percent is absorbed by the surface. It is emitted as heat (2). This has difficulty in escaping because it is absorbed by gases in the Earth's atmosphere, so increasing the temperature near the Earth's surface.

across the Earth's surface. As continental drift alters this distribution, so the effects will change. For example, when the continents are gathered into one supercontinent, the Milankovitch variations should produce greater temperature fluctuations, with a higher maximum temperature, than at present.

The Earth and the Moon

The gravitational influence of the Moon on the Earth actually stabilizes the direction of the Earth's axis in space. So the Milankovitch variations are kept fairly small. This will not necessarily be true in the future. The stability depends on the Moon's interaction with the rotation of the Earth. The Moon is involved (along with the Sun) in raising tides in the Earth's oceans. These can be thought of, in simple terms, as oceanic bulges on either side of the Earth—pointing toward and away from the Moon—which are produced by the Moon's gravitational effect. The Moon moves round the Earth much more slowly than the Earth rotates (one month as compared with one day). This means that the solid Earth rotates under the tidal bulges. The result from our viewpoint is that we see two high tides a day as the bulges pass us by. The Moon is trying to hold the oceans back, but the Earth is trying to move them forward. This creates friction between the oceans and the solid Earth. The result of the tug-of-war is that the Earth is made to spin a little more slowly. At the same time, the Moon is forced to circle the Earth at a slightly greater distance.

Such *tidal friction* is created especially in the shallow seas at the edges of continents. Since the extent of the shallow seas varies at different stages in the supercontinental cycle, the amount of friction must change with time. This means that the rate at which the Earth's rotation slows down should also vary. One estimate is that the slowing down has varied by a factor of two over the past 450 million years. The length of a day is currently increasing by about two-thousandths of a second per century. This means, in turn, that the Moon is moving away from the Earth by about 4 meters per century. These amounts sound very small, but the forces at work continue acting for long periods of time. Over the next billion years, the Moon's orbit will increase in size by tens of thousands of kilometers.

The movement of the Moon away from the Earth has a minor, but interesting consequence for the far future. One of the more spectacular sights that astronomy has to offer is a total eclipse of the Sun. The Moon is currently at just such a distance from the Earth that, when it eclipses the Sun, it can block out the dazzling light from the surface of the Sun. This allows us to see briefly the extensive, but faint, solar atmosphere. As the Moon continues to move away from the Earth, its apparent size in the sky will decrease. Instead of entirely blacking out the Sun's disk, it will leave part of the disk always showing. That will mark the end of spectacular solar eclipses for us on Earth, since the light from this exposed part of the surface will prevent us from seeing the faint solar atmosphere. Because the rate at which the Moon leaves us decreases the further away it is from the Earth, it will be a long time before this comes to pass. Whether anyone will be around then on the Earth's surface to regret the passing of total solar eclipses is dubious.

As it happens, much of our knowledge of the slowing down of the Earth's rotation depends on observations of solar eclipses. Babylonian astronomers long ago kept records of such eclipses. The clay tablets containing their observations have been dug up and their information analyzed over the past century. The striking result is that the times they record for the eclipses differ from what we would calculate today. The difference is due to the slowing down of the Earth. By using these records, dating back some two to three millennia, the rate of slowing down of the Earth's spin over this period of time can be calculated.

The Sun and the Moon are about equally effective in producing tides in the Earth's oceans. This is because the Sun, though much further away, is also much more massive than the Moon. The two sometimes work together, and sometimes in opposition, as the Moon orbits round the Earth. Consequently, the actual height of high tide varies during the month. What is less obvious is that the Sun and the Moon also raise tides in the solid body of the Earth (in other words, they slightly distort its shape). These tides are less effective at changing things than the ocean tides, but they cannot be totally ignored. The Moon is famous for keeping its same face toward the Earth all the time. (This is why we always see the "Man in the Moon.") The Moon has never had oceans: it is tides, caused by the Earth, in the solid body of the Moon that have slowed its rotation until it keeps the same face toward us. Tides occur widely throughout the solar system, though their effects are often small enough to be ignored.

In the end, tidal friction will only stop operating when the Earth rotates once in the same time that it takes for the Moon to go once round the Earth. When this happens, the tidal bulge will always remain over the same place on the Earth's surface, as also will the Moon. (So there will be no more Moon-driven tides of the sort we know today.) This means that the balance point is reached when the day has the same length as the month. Because tidal interaction makes the Moon move into a larger orbit, it will go round the Earth more slowly as time passes. Consequently, the month is gradually getting longer. A stable position will be reached when the month contains about forty-seven of our present-length days. The length of a "day" on Earth will then also be forty-seven days. In addition—to return to our original point—a more distant Moon will be less able to stabilize the direction of the Earth's axis in space. So the Milankovitch variations will become much larger: the Earth's axis will nod backward and forward to a far greater extent than at present. The Sun's heat will then be smeared out over different latitudes, sometimes more and sometimes less than at present. This will act to produce greater variations in climate. Some calculations suggest that, within the next 4.5 billion years, the Earth's axis might even dip through nearly 90°. Instead of the Sun being overhead at the equator, it would be overhead at the poles. Changes of this sort will obviously greatly alter the distribution of temperature across the Earth's surface.

Gases in the Atmosphere

Some of the things that can significantly affect climate are less regular in their action than tidal friction. One example is huge impacts on the Earth's surface—which will be discussed in a later chapter. Another is volcanic activity. In principle, this is related to the pattern of continental drift. For example, one place where volcanoes typically occur is at the edge of a continental plate, where an ocean plate dips below it. As the ocean plate moves down into regions of higher temperature, it produces gases and hot rock which expand and burst upward through the Earth's surface, forming a volcano. The gases are vented into the atmosphere along with a quantity of dust. Dust from a powerful volcano can pollute the whole atmosphere, cutting off the sunlight and so leading to a fall in temperature at the surface of the Earth. Prolonged periods of high volcanic activity can therefore act as a trigger for climatic change.

The gases vented by volcanoes typically have water—in the form of steam—and carbon dioxide as their commonest constituents. This is hardly surprising since ocean sediments inevitably contain some water, and often include tiny shells which have fallen to the ocean floor when their owners died. These shells are typically made of minerals which give off carbon dioxide when heated. Consequently, the effect of continental drift is to recycle the more volatile materials deposited on the ocean floors. The steam given off soon condenses into water, and ultimately finds its way back to the oceans. The fate of the carbon dioxide is more complicated. Some is absorbed by plants; some dissolves in the oceans; some remains in the atmosphere. The remnant in the atmosphere has an important effect on the climate. Carbon dioxide readily absorbs heat radiation, and so acts, along with water vapor (another good heat absorber), to keep the heat from the Sun confined near the Earth's surface, rather than letting it escape rapidly into space. In other words, these two gases (and some other less common ones) provide an insulating blanket that keeps the Earth warmer than it would otherwise be. The gases are said to produce a *greenhouse effect*: so called because the glass in a greenhouse is good at letting the sunlight in, but only lets heat leak out slowly. The greenhouse effect is vital for living things on the Earth. The average surface temperature of the Earth is currently about 15 °C. Without the greenhouse effect, it would be some 30 °C lower. At an average temperature of 15 °C below freezing, much of the water on Earth would become ice, and anything except very simple life would be hard hit.

Too little carbon dioxide in the atmosphere, and the temperature of the Earth's surface would drop. But, equally, there are

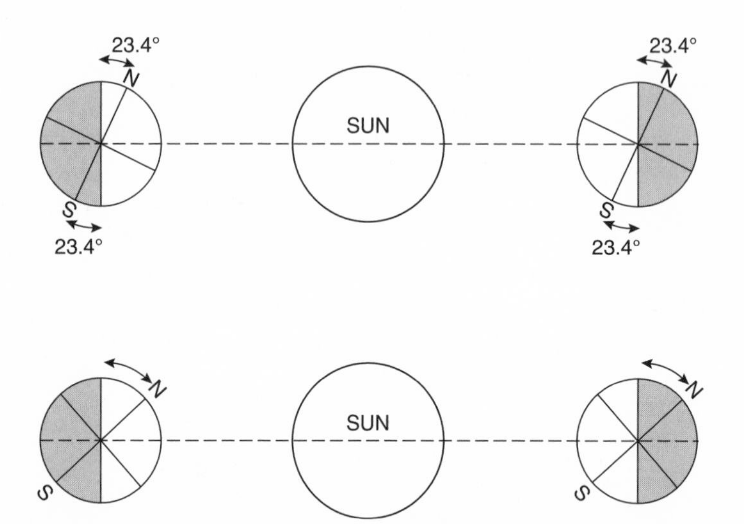

3.3 The Milankovitch effect. The top diagram shows the situation today, where only small wobbles of the Earth's axis occur. The lower diagram shows the much larger wobbles that can happen when the Moon is far away from the Earth in the future.

problems with too much carbon dioxide. It could raise the temperature of the surface to uncomfortable levels, as is stressed in the current debate about a manmade greenhouse effect (see below). The important thing for the Earth's future is the long-term balance of carbon dioxide, and this depends on the recycling processes at work. When carbon dioxide in the atmosphere dissolves in water, it makes the water slightly acid. If this water then runs over a rock the acid will dissolve some of the rock and carry it away. Most of this dissolved rock is carried down by streams and rivers to end up in the sea. It has been estimated that a land mass 1–3 kilometers high could be reduced to sea level by this process over a period of 50–100 million years. This is the same sort of timescale as the periods involved in continental drift. In fact, there is a continual conflict between continental drift, which elevates land masses, and weathering processes, which erode them away.

Since mountain building varies with the supercontinental cycle, the rate of removal of carbon dioxide from the atmosphere via weathering should vary with the same kind of period. Removal of significant amounts of carbon dioxide decreases the atmospheric greenhouse effect, and can lead to an ice age. Looking backward in time, there is some evidence that ice ages have indeed recurred on a timescale of 100–200 million years, similar to the timescale for the growth and decay of a supercontinent. Before human activities affected our planet, we were in an interglacial period that had perhaps another 10,000 years to run. It is unlikely that manmade warming can do more than extend this period slightly. Then the Earth will lapse back into an ice age again. In the long term, the importance of weathering is that some of the Earth's carbon dioxide is locked away permanently, not to return to the atmosphere again. Consequently, the total amount of carbon dioxide in the atmosphere has declined over the past 2–3 billion years (though more so in the earlier years than recently). Currently, most carbon dioxide (90 percent) is combined with other materials to form solid rocks. Much of the rest is dissolved in the oceans, with only a trace in the atmosphere.

Carbon dioxide is not the only gas whose presence in the atmosphere has changed with time. The geological record suggests that oxygen only began to enter the atmosphere in quantity 2 billion years ago. It has accumulated as a by-product of the growth of life on Earth. Nitrogen too has entered the atmosphere mainly as a result of living processes. In fact, it has been suggested that the best way of detecting life on other planets is to look for unexpected gases in their atmospheres. The Earth is the only planet in the solar system whose atmosphere is dominated by nitrogen and oxygen. It is also, so far as we know, the only planet with life. Though the amounts of nitrogen and oxygen in the atmosphere have increased in the past, they do not appear to be changing much with time now.

It is generally accepted that life on the Earth's surface plays an important role in regulating atmospheric change. Suppose the amount of carbon dioxide in the atmosphere increases. Plants use carbon dioxide to build themselves up. Give them more carbon dioxide, and they will use it to build themselves up even more. To put it another way, increasing the amount of carbon dioxide in the atmosphere increases the rate at which photosynthesis occurs. This means, in turn, that the amount of carbon dioxide in the atmosphere decreases. The opposite occurs if the amount of carbon dioxide decreases: the rate of photosynthesis goes down, so more carbon dioxide remains in the atmosphere. What is happening is that a change in the amount of carbon dioxide leads to feedback that opposes the change. The Gaia hypothesis takes this idea of feedback to the limit. (It takes its name from Gaia, the Greek goddess of the Earth.) The hypothesis suggests that life on Earth actually regulates conditions here in such a way as to make sure that living processes can continue in the future. Even so, it is generally agreed that such feedback can break down if conditions change too rapidly or too much.

The Future of the Atmosphere and Oceans

Which brings us to the question of what the future has in store. One lot of changes relates to the mobility of the Earth's surface. We saw in the previous chapter that, as the Earth cools, the movements in the mantle will slow down and eventually stop. This means that continental drift will do the same. Ocean currents will therefore change their basic pattern less frequently, eventually stabilizing to fit whatever final configuration is reached by the continents and oceans. The world will be warmer or colder depending on how freely equatorial currents can flow as this end layout is approached. The winds too will adapt to the final continental pattern, but again we have to remember the weathering process. Uplift of the continents depends primarily on the motions of the surface plates. If this slows down and stops, weathering will race ahead of mountain building. The continents will become low-lying areas. The winds will blow more steadily across them, and changes due to different heights of land will be minimal. The exceptions will be places where plumes continue to rise slowly to the Earth's surface from the cooling core. Here uplifts of the land can still occur.

The slowing down of the Earth's rotation leads to a reduction in the Coriolis force. The winds (and ocean currents too) will therefore move increasingly north–south, rather than northeast–southwest. This makes the interchange of heat between the equator and the poles easier, but will not necessarily alter the present system of three basic atmospheric cells from low to high latitudes. The existence of the cells depends on the rate at which hot air radiates its energy to space and then sinks back to the surface. But another factor enters here. Currently, there is not much difference in temperature between the day and night sides of the Earth because nights are fairly short. Before the night side has had time to radiate much heat to space, the Earth has turned, and the dark side has become the day side again. A slow-down in the rotation rate as the Moon retreats means that the night side has much more time to lose heat. The larger differences in temperature between the day and night sides that this produces leads to an increased flow of air between the two sides. So, along with the north–south winds, there are winds blowing from the day side to the night side. For any particular spot on the Earth's surface, the wind pattern will vary with the length of the month, as the Earth slowly presents its different faces to the Sun. The pattern will change drastically if the Earth's spin axis experiences a major tilt, as it has a fair chance of doing when the Moon retreats. For example, if the north pole pointed toward the Sun, the atmospheric circulation would involve air currents from the north pole to the south pole.

The immediate uncertainty relates to the possible heating of the Earth's surface due to a manmade increase in the greenhouse effect. The activities of human beings have been affecting the atmosphere for a long time, but it is only in the last two centuries, as the burning of fossil fuels has drastically increased, that the effects have become noticeable. The main gas produced is carbon dioxide, and this has been piling up in the atmosphere. There is a general feeling that the increased concentration has already led to a slight rise in global temperature. But the important question concerns what will happen in the future. If carbon dioxide continues to be pumped into the atmosphere, as at present, for the whole of the coming century, then the average temperature in 2100 could be somewhere between 2 and 5 °C above its current value. This does not sound very much, but its consequences would be profound. Cities round the globe which currently have acceptable summer temperatures would bake. Places like Dallas, Las Vegas, Jerusalem and Athens might expect temperatures of 40 °C or more. The hotter climate at these latitudes would be correspondingly drier, threatening water supplies that in many areas are already stretched. For others, the problem would be too much water. The higher temperature would melt much of the ice in glaciers and in the polar caps. The water released, plus the expansion of the oceans due to the higher temperature, would raise the sea level. This could drown low-lying islands in the Pacific, or low-lying areas of continents from Bangladesh to the Netherlands. At the same time, the higher temperatures could well lead to more violent hurricanes. So an area such as Florida would be at risk from these, as well as from higher temperatures and flooding.

Of course, there will be winners as well as losers. Countries at higher latitudes are likely to gain more acceptable climates than those they currently enjoy, and so perhaps become easier places in which to live. There is no question of the human race dying out; the problem is rather how to deal with the immense problems that the forecast changes will bring. It has been estimated, for example, that tens of millions of Bangladeshis will be displaced by the predicted flooding. It is not obvious where they could go. This picture of future events depends, of course, on the accuracy of the theoretical predictions for the future. Because of the complexity of the Earth's atmosphere, actual future happenings may be better, or worse. At least the standard scenario leaves a little time for preparation. As against this, one worry has been that slow warming might, at some point, trigger off a rapid major change. It has been suggested, for example, that the Antarctic icecap might, when it is warmed to a certain level, disintegrate very quickly. Again there have been worries about methane deposits buried below the oceans. Methane, as a gas, is considerably more efficient than carbon dioxide in terms of producing a greenhouse effect. Fortunately, it is also much less commonly found. It has been suggested that the warming of the oceans might at some point release large quantities of methane in a relatively short time. If it did, the average global temperature would jump rapidly. It is difficult to disprove such suggestions, but the general belief is that they are unlikely to be important in the next hundred years. The standard scenario is the one to worry about for the coming century.

From the viewpoint of the Earth's future, manmade warming can best be seen as a short-term hiccup. Many species may be unable to cope with the rapid changes, but disastrous loss of life has occurred before on Earth. Recovery will happen, as it has happened before. Human communities will be under stress for some centuries, but longer-term consequences are likely to be limited. One aspect of life that may be affected is the onset of the next ice age. Judging by past ice ages, the global average temperature should soon start on a gradual decline. Manmade warming may defer this, so that we enjoy a prolonged interglacial. But the long-term trend is clearly downward. The gradual loss of carbon dioxide from the atmosphere as rocks weather, means that the greenhouse effect should decrease slowly with time. This means, in turn, that ice ages should become more probable in the future. There is another consequence. Plants need carbon dioxide to live. If the level of carbon dioxide in the atmosphere falls, a stage will eventually be reached when complex plant life on Earth can no longer survive. It is estimated that this will happen before the end of the next billion years. Without plants, the amount of oxygen in the Earth's atmosphere will decrease. In fact, it could well fall drastically over a period of only 10 million years after their disappearance. And, of course, no oxygen implies no advanced animal life on the Earth's surface. But will this scenario happen? Other future changes could bring about a similar end-result, but by a different route.

All the developments that we have looked at so far become important over periods of one to a few billion years. But there is another mechanism operating on the same timescale which is more significant than any of these—the gradually increasing heat from the Sun. We saw in Chapter 1 that, as the Sun consumes the hydrogen near its center, so it gradually becomes brighter and hotter. The increased heat from the Sun means that the Earth's surface will also become hotter. An important turning point will come when the Earth's surface becomes hot enough for the oceans to boil, turning into water vapor in the atmosphere. A simple calculation suggests that it will only occur many billion years in the future. But this does not allow for the greenhouse effect. As the surface temperature of the Earth increases, the amount of water vapor entering the atmosphere will increase in parallel. Water vapor is a good heat absorber, so its presence will lead to a greater greenhouse effect. This produces a still higher surface temperature, which brings even more water vapor into the atmosphere. If the Earth can get rid of this water vapor to space, then the temperature will increase, but fairly slowly. The rise will be offset, in part, by loss of carbon dioxide due to weathering. The problem comes if—as seems likely—the water vapor accumulates in the atmosphere. At a certain point, the feedback process will then take off rapidly, producing a "runaway" greenhouse effect. When this happens, the surface temperature will keep on rising until all the water has entered the atmosphere (which is why the word "runaway" is used). A runaway greenhouse effect is still some time away, but could perhaps happen in 1–2 billion years from now. Surface temperatures could then rise to 1,000 °C, at which point some of the surface rocks would begin to melt. The Scottish poet, Robert Burns, told his sweetheart that his love would endure: "Till all the seas gang dry, my dear, And the rocks melt wi' the sun." Current estimates of the time this will take would surely have satisfied the most demanding sweetheart.

As the temperature rises, all except the simplest forms of life on land will disappear. This should happen by about a billion years from now, when the surface temperature will have reached around 70 °C. Life in the oceans will continue longer, but will also eventually be killed off. As we have seen, the composition of the atmosphere is related to the existence of life on the Earth's surface. If life ceases, the oxygen and nitrogen in the atmosphere will both combine chemically with the surface, and so disappear. Oxygen in the upper atmosphere—in the form of ozone—has a significant effect in protecting the Earth's surface from ultraviolet light. However, in the absence of life on the surface, the only things to suffer an excessive suntan will be rocks. The loss of the oceans means that continental drift, which has already been slowing down, will stop. When the oceans disappear, the sediments on their floors will be directly exposed to the increasing heat. They will give up some of the carbon dioxide they have captured to the atmosphere, increasing the greenhouse effect even further.

Our picture of the Earth as it will be some 2–3 billion years in the future is of a declining world. The solid surface is mostly inactive, the temperature is rising inexorably, and life has disappeared. In fact the Earth will look rather like Venus does at present—mostly inactive and very hot. But this rather unattractive picture is not the final stage. The Sun will continue to brighten. As it heats the Earth more and more, much of the Earth's atmosphere will be lost to space. Increasingly, surface rocks will be heated, melted, and even—for some of the lighter elements—evaporated to form a thin "metallic" atmosphere round the Earth. One intriguing possibility is that the Earth will by now be keeping the same face pointed toward the Sun all the time. The tug of the Earth has produced tidal effects in the solid Moon with the result that we always see the same face. In a similar way, tides produced by the Sun, acting over a long enough period, may eventually slow the spin of the Earth so that it always keeps the same face turned toward the Sun. If this happens, the face pointing toward the Sun will become very hot, while the side pointing away will become very cold: the temperature there may fall as low as –240 °C. Winds blow from hotter to colder places. The material melted from rocks on the hot side of the earth will circulate round to the cold side, where it will condense into exotic kinds of "snow" and fall back to the surface.

Meanwhile, the Sun continues to grow in size. Its final evolution, turning it into a red giant, will make it expand out to just about as far as the Earth's present orbit. This will happen some 8 billion years from now. Yet even this expansion does not necessarily spell the end of the Earth. As we saw in Chapter 1, when the Sun expands, it will blow off a considerable part of its mass into space. But it is the Sun's gravitational pull that keeps the Earth in its present orbit. Decreasing the amount of material in the Sun decreases its gravitational pull. The result is that the Earth will move into a more distant orbit. By the time the Sun expands to the Earth's present orbit, the Earth itself will have moved out to nearly twice its present distance from the Sun. This is much too far away for it to be gobbled up directly by the Sun. Yet the Earth and its environment will have changed greatly by then. For example, the Earth will have been plowing its way through the large amount of material shot off from the Sun for some time past. These environmental changes may cause the Earth to spiral back toward the Sun. So, even if we avoid being consumed when our central star expands, it may still get us in the end. The Moon, already remote from the Earth, stands a good chance of having left us altogether by then, wandering away as an independent body.

Even if the Earth survives all these perils, it will be as a far more boring planet than it is today—little atmosphere, no oceans, no surface activity, and with life long since departed. Supposing that it survives all the changes in its environment as the Sun evolves from a red giant to a planetary nebula, it will continue its existence as a dead planet. In this state, it will accompany a faint and dying Sun—now a white dwarf star—into the far distant future. If this prophecy sounds rather gloomy, maybe we should cheer ourselves with a comment by Dr. Samuel Johnson. He was offered a remarkably similar picture of the future by eighteenth-century scientists. His comment was: "the ocean and the Sun will last our time, and we may leave posterity to shift for themselves."

4. Magnetic Earth and Magnetic Sun

The Source of the Earth's Magnetism

Magnetic compasses were in use for many centuries before it was finally guessed how they worked. The breakthrough came with William Gilbert, physician to Queen Elizabeth I, who showed that compasses point in a particular direction on the Earth's surface because the Earth itself is a great magnet. By the beginning of the seventeenth century people knew that pieces of magnetized iron could affect which way nearby compass needles pointed. The needles always point toward one end of the magnet and away from the other end. This is very similar to what happens on the Earth's surface, where compasses always orientate themselves in a more or less north–south direction. (By analogy with the Earth, magnets are therefore said to have north and south poles.) Gilbert suggested that the Earth's interior was essentially a large magnet, lined up roughly north–south, and this was what forced compass needles at the surface to line up in the same direction.

As time passed, this idea of the Earth as simply being a magnetized ball raised an increasing number of problems. It was found, for example, that, though the magnetic poles remain north and south, they do tend to wander about a bit. (They are currently some 11° of latitude away from the geographical north and south poles.) How could this movement happen in a solid Earth? The need for a complete rethink finally became clear when the internal structure of the Earth was established (see Chapter 2). This showed that the Earth did indeed have an iron core, but, because it was molten, there was no way that such a core could act as a magnetized iron bar. So arose the key question: what creates the Earth's magnetism?

The beginnings of our modern explanation go back to Michael Faraday, in the first half of the nineteenth century, and to his research into *electromagnetism*. He coined this compound word because his work showed that electricity and magnetism were not separate things as had previously been supposed. Electricity could produce magnetic effects, and vice versa. He described this interaction in terms of what happens in the region round a magnet. He called this space a *field*, imagining that every magnet has around it an invisible magnetic field. His picture is a little like that of a cricket or baseball game. The essential part of the game is the narrow strip that joins the batsman to the bowler, or the pitcher to the hitter. But what happens there affects other players spread out over the surrounding area, and is in turn affected by them. It is the combination of the action at the center and the activities in the field round about that determines the result of a game. In a similar way, a magnet may be a clearly defined narrow strip, but it affects, and is affected by, what is happening in the region around it.

One of the things that resulted from Faraday's picture was the invention of the dynamo. He found that spinning a metal disk in the presence of a magnetic field produced electrical currents in the disk. This is the essence of a dynamo. It uses motion (in this case of a metal disk) to convert magnetism into electricity. Later on a further twist was added to the picture. The electricity generated in the disk can be led off and passed through a coil of wire. This makes the coil produce a magnetic field around itself. If the coil is positioned close to the disk, the latter will then be spinning in the magnetic field of the coil. This leads to the following sequence. The coil produces a magnetic field which the spinning disk converts into electrical currents. These currents are fed back into the coil to produce more magnetism; this, in turn, produces more currents in the disk, and so on. Such a setup is called a *self-exciting dynamo*. So long as the motion continues, so too will the production of magnetism and electricity. Both the Earth and the Sun are magnetic, and their magnetism is believed to be due to self-exciting dynamos in their interiors.

It is difficult, at first sight, to see the link between a dynamo consisting of solid parts, on the one hand, and the fluid core of the

Earth, on the other. So how does it work? The secret is that the Earth's core actually has the basic requirements needed for dynamo activity. Three things are required: an electrical conductor, appropriate motions of this conductor, and an initial magnetic field to start things off. Molten iron, the main constituent of the Earth's core, is a good conductor of electricity, and all of it—as we have seen—is in motion. In fact, it has two sets of motion—one up and down due to convection, and the other round and round due to the Earth's rotation. Moreover, there was some magnetism around at the beginning of the solar system which both the Earth and the Sun picked up. Long and laborious computer calculations have shown that this initial magnetism, together with the two sorts of motion found in the Earth's core, are just what is needed to produce a self-exciting dynamo. The strong electrical currents circulating in the core due to this dynamo produce a strong magnetic field down there. Up at the surface, we only observe a much weaker field, but it is quite sufficient to affect ordinary compasses.

Though a self-exciting dynamo is much more complicated than a simple bar magnet, it fits the observations much better. For example, the Earth's magnetic field is not as regular or symmetrical as the field of a bar magnet. The deviations can be explained by minor swirls in the fluid of the core, probably due to small bumps and hollows on the inner surface of the mantle. So the rough mantle boundary complicates the motions in the core. (This is offset to some extent by the influence of the solid inner core, which seems to smooth things out.) As mentioned in Chapter 2, rocks forming in the presence of the Earth's magnetic field may retain an imprint of the magnetism present at the time they formed. Some of the oldest rocks on Earth have retained such an imprint, confirming that the dynamo in the core was already at work in the early days of the Earth. Indeed, since the Earth's interior was hotter then, and it was spinning more rapidly, the dynamo probably worked more efficiently than now.

But the rock measurements also suggest surprising fluctuations in the Earth's field. It seems that now and again the Earth's magnetism has flipped: sometimes our present north magnetic pole has been a south pole, and vice versa. These magnetic flips are actually predicted by theory; but, unfortunately, current theories cannot say when precisely they will occur. Periods of a few million years between such magnetic reversals are common, but sometimes they are shorter and sometimes longer. The last reversal happened just over three-quarters of a million years ago (prior to that, a compass would have pointed southward), but there have been periods of up to 40 million years in the past that have seen no reversal. The problem is that quite small changes in the fluid currents in the core can trigger such reversals. From our vantage (or possibly disadvantage) point on the Earth's surface, the time between magnetic flips looks fairly random.

The Future of the Earth's Magnetism

Looking to the future, the immediate interest relates to these magnetic reversals. In particular, when will the next one occur? The actual process of reversing the Earth's magnetism takes a relatively short time – only a few thousand years. This is much shorter than the million years or more during which the direction of the Earth's magnetism, whether it points north or south, remains the same. Consequently, the probability of living through a reversal is fairly small. During a reversal, the magnetic field of the Earth both gets more complicated, and also weakens. Interestingly, observations of the magnetic field over the past few centuries suggest that it has been getting more complex. For example, there is a large area round South Africa where the magnetic field is now pointing in the opposite direction to that observed in surrounding areas. Moreover, the Earth's magnetism has decreased appreciably in strength over the last few hundred years. All this suggests that, despite the odds, we may be coming up to another flip in its direction some time during the next few thousand years.

In the longer term, what the Earth's magnetism looks like at the surface will depend on the conditions at the two boundaries—between the mantle and the liquid core, and between the liquid core and the solid inner core. As we have seen, convection currents in the mantle circulate on a timescale of around a hundred

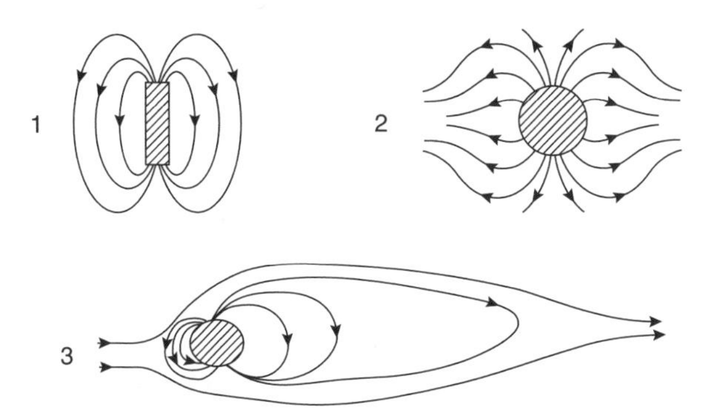

4.1 A comparison of magnetic fields. 1 the magnetic field round a bar magnet; 2 the Sun's magnetic field (notice how the solar wind pulls it out round the Sun's equator); 3 the Earth's magnetic field (the solar wind, coming from the left, pushes the field in on the sunward side, and pulls it out on the opposite side).

million years. Their up-and-down motions are likely to alter the surface of the mantle where it meets the core. As the bumps and hollows change, so the detailed pattern of the magnetic field at the Earth's surface will slowly change in sympathy. On a still longer timescale, the growth of the inner core will introduce change. It should lead, in due course, to greater stability for the dynamo process, making magnetic reversals more difficult. On a timescale of a billion years or more, the growth of the solid core will decrease the space available for the liquid core. Simultaneously, the Earth's rotation will be decreasing due to its interaction with the Moon. Both of these changes can affect the efficient operation of the Earth's dynamo. Calculations suggest that, even so, the dynamo should continue to work, but the field produced will be weaker. Finally, as the Earth continues to cool, the dynamo action stops, and the Earth loses its magnetism.

The Source of the Sun's Magnetism

One of the first discoveries that Galileo made, after his construction of an astronomical telescope, was the presence of dark patches on the Sun's surface. The comings and goings of these patches—now labeled *sunspots*—have been followed for some four hundred years. Almost a century ago, it was found that sunspots are highly magnetic. They typically appear in pairs, oriented roughly parallel to the Sun's equator. One member of the pair corresponds to a north magnetic pole, the other to a south pole. Some time later, it was found that the Sun also has a much weaker general magnetic field. This is rather similar to the Earth's field, with north and south magnetic poles near to the poles of the Sun's spin axis.

This description makes the Sun's magnetism sound distinctly different from the Earth's. Yet it can, in fact, be explained in exactly the same way—as due to a self-exciting dynamo. The apparent difference simply reflects the different internal structure of the Sun. As explained in Chapter 1, the outer layers of the Sun are convective, while the inner parts are not. Since this convective zone is at a high temperature, the atoms in it are broken up into charged particles. Such particles can conduct electricity. Along with convection, the outer layers of the Sun are rotating. Here we have the conditions required for a self-exciting dynamo to work just as on Earth—electrically conducting material, convection, and rotation. The difference between the Earth and the Sun is that the conditions apply to the central parts of the Earth, but to the outer parts of the Sun. Observations suggest that the dynamo process in the Sun is concentrated around a region about a quarter of the way into the Sun. This is where the transition from the inner core to the outer layers particularly churns up the solar material.

Both the Earth and the Sun have north–south magnetic fields. The difference between the two bodies is the appearance of sunspot fields on the Sun. On Earth, the strong magnetic fields generated in the core do not reach the surface. All we see is the weaker north–south field. On the Sun, the strong magnetic fields are generated closer to the surface. Loops of these strong fields can therefore float up to the surface. Where they intersect it, sunspots are formed. Consequently, an observer can see both types of magnetism on the Sun, but only one type on the Earth. Like the Earth, the Sun changes the direction of its magnetism from time to time. But the changes on the Sun occur more regularly, and much more frequently, than those on Earth. Both the general magnetic field of the Sun and the magnetism of the sunspots flip direction every 11 years or so. (This period is usually called the *solar cycle.*) For the general field, this means that the magnetic north pole becomes a south pole and vice versa, as on Earth. It is a little more complicated for sunspots. What happens to them depends on the fact that they occur in pairs. Each pair is carried around by the Sun's rotation, with one spot leading and the other following. Suppose that the leading spot of a pair in one cycle is a north pole, and the following spot a south pole. Then in the next cycle, this will be reversed: the south spot will lead and the north spot will follow.

Variations in the Sun's magnetism are reflected by the changing number of sunspots visible during the course of a solar cycle. There have been extended periods when sunspots have been hard to find. For example, sunspots were rare during the period from 1645 to 1715. (This period is now called the *Maunder minimum,* after the British astronomer who discussed it.) The strong magnetic fields of spots have a major influence on the Sun's atmosphere above them. They both heat it up and make parts of it move rapidly, sometimes explosively. For this reason, changes in the magnetic fields threading through the Sun's surface and atmosphere are often referred to as changes in the Sun's *activity.* The Maunder minimum is therefore properly described as a time of low solar activity.

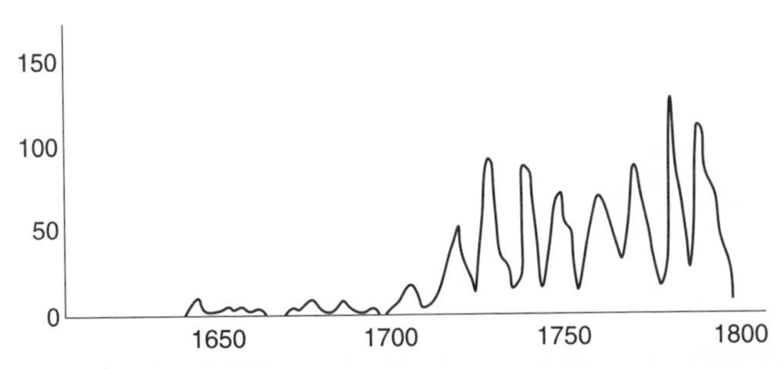

4.2 The Maunder minimum. Very few sunspots were seen in the latter half of the seventeenth century, but they reappeared in the eighteenth century. Minima of this sort can be expected to recur at intervals in the future. The number of sunspots visible each year is shown vertically, and the year concerned is shown horizontally.

The Future of the Sun's Magnetism

In trying to forecast how the Sun's magnetism will develop in the future, we can draw on information of a type that is not available for the Earth. The Sun is a common sort of star: there are plenty of others like it in our vicinity. They differ mainly in terms of age: some are older than the Sun and some younger. Though we cannot wait around to see how the Sun's magnetism will develop in the distant future, we can get an idea of what will happen by looking at the magnetic properties of Sun-like stars which are older than the Sun. It turns out that it is a good deal easier to measure the activity of a star—that is, the disturbances in its atmosphere—than to make a direct measurement of its magnetism. So long as this activity is always related to the stars' magnetism, we can use such observations just as well to work out how the magnetism of the stars varies with age. The result of this comparison is clear—the magnetic activity of the Sun will decrease with time. The early Sun must have been very active, with larger and more irregular outbursts than at present. This initial display of magnetic power declined fairly rapidly at first, and then more slowly as the Sun aged. This slow decline will continue while the Sun remains on the main sequence. Toward the end of its main-sequence lifetime, the Sun's magnetic activity may be a quarter or more down on its present level.

Related observations also make it possible to measure the spin rates of stars like the Sun. When such stars are arranged in an age sequence, it is clear that the younger ones spin much faster than the older ones. Subsequently, they divide into two groups: about a half continue to spin rapidly, while the remainder slow down. The Sun clearly belongs to the latter group. Not long after its birth, the Sun would have been spinning much faster than at present—perhaps one rotation every five days. The rate then decreased quite quickly—by astronomical standards. Half-way through its lifespan from birth to the present, its rotation period had already slowed down to 20 days. It is currently about 26 days. By the time the Sun reaches the end of its main-sequence lifetime, its spin rate will have dropped a little further—to around 30 days. Since the Sun's dynamo depends in part on the rate of rotation, this slowdown is presumably linked to the decrease in the Sun's activity. But why is the Sun slowing down?

The Solar Wind

The magnetic activity at the Sun's surface interacts with the electrically charged particles in its atmosphere. It makes them move about more quickly, which is equivalent to saying that it heats the atmosphere. The result is that the outer layers of the Sun's atmosphere continually expand outward into interplanetary space. This flow of gas away from the Sun is called the *solar wind*. The particles in the wind are still electrically charged and so continue to interact with the Sun's magnetism as it stretches out past the planets. Indeed, the Sun's magnetic field, as it is carried round by the Sun's rotation, tries to keep the particles moving with it. The particles, meanwhile, are trying to move straight outward away from the Sun. Their opposition to being pulled in another direction—round the Sun—slows down the spin of the Sun. Think of a skater who is spinning round with her arms by her side. This is the situation when the particles are at the Sun's surface. If the skater now extends her arms out sideways, her spin rate drops appreciably. The Sun, in trying to make the distant particles spin with it, is acting like a skater who spreads out her arms.

The spin decreases only slowly with time because the solar wind contains much less material than the Sun. (It is like a very fat skater with tiny arms.) If the solar wind contained much more material, the spin of the Sun would decrease more rapidly. In the early days of the Sun, solar activity was stronger, so more material flowed away from the Sun. This stronger solar wind slowed down the Sun's spin more efficiently. But an interesting feedback now occurred. As the Sun began to rotate more slowly, it diminished the solar dynamo, which meant that magnetic activity at the solar surface decreased. As a result, the braking effect of the solar wind on the Sun's rotation worked less well, and the falloff in the spin rate became slower. This meant, in turn, that the decrease in magnetic activity slowed. So the Sun's rotation and its magnetic activity have tailed off together. They will continue to do so, though at a decreasing rate, while the Sun remains on the main sequence.

Since the strength of the solar wind depends on the amount of magnetic activity on the Sun, it varies with the solar cycle. In fact, the strength and gustiness of the wind can be related to the area of the Sun's surface covered with spots. During periods with few sunspots—the Maunder minimum, for example—the solar wind becomes much weaker. It has been estimated that such periods may occur for a third of the time, with the ordinary solar cycle operating for the other two-thirds. In the distant future, as the Sun's magnetism declines, such extended sunspot-free periods will become increasingly common. Strong solar winds will correspondingly occur less frequently.

As the Earth moves round the Sun, it is continually plowing its way through the solar wind. The wind, since it consists of electrically charged particles, interacts with the Earth's magnetic field. The pressure the particles exert compresses the Earth's field on the side facing the Sun. On the other side, the wind blows the magnetic field away into a lengthy tail. The whole volume occupied by the Earth's field—the equivalent of the whole of the playing area in cricket or baseball—is called the *magnetosphere*. (To call it a "sphere" is not very helpful: it looks more like a comet with its tail pointing away from the Sun.) The magnetosphere acts as a barrier to the solar wind, so that most of the particles from the Sun simply flow past the Earth on either side. But the magnetosphere has two holes—at the north and south magnetic poles. Here particles from the solar wind can filter down to the Earth's atmosphere. When the Sun is particularly active, the solar wind contains frequent strong gusts of particles. If the Earth runs into one of these gusts, the upper levels of the Earth's atmosphere are peppered with these solar particles. This produces the lights that we call an *aurora*. We can expect that, in the future, as the solar wind becomes less gusty, great auroras will become less frequent.

The solar cycle has another influence on the Earth. The Earth's surface seems to be slightly warmer when the Sun is particularly active. The reason is not well understood, but the solar wind may play a role here too. The average increase is small—only a fraction of a degree—but, like the Milankovitch variations described in the previous chapter, it can have a significant effect on the Earth's climate. For example, winters were especially harsh during the Maunder minimum. The long-term decrease in solar activity should therefore make the Earth colder. However, this relatively small effect will be swamped by the parallel increase in the amount of heat coming from the Sun.

Losing an Atmosphere

When a space probe is launched, it must reach a certain minimum speed—called the *escape velocity*—to get away from the Earth's gravitational pull. This minimum speed varies from planet to planet, depending on how massive the planet is. For the Earth, the escape velocity is 11.2 kilometers per second. Anything that is to leave the Earth must acquire a speed greater than this. It follows that an atom in the upper atmosphere of the Earth can also get away, if it can reach this kind of speed. How fast the atoms and molecules in the Earth's atmosphere move depends on two things—the temperature of the atmosphere and how massive the particle is. Higher temperatures and lower masses produce higher speeds. A study of the top layers of the Earth's atmosphere shows that atoms of the two lightest elements, hydrogen and helium, can sometimes reach speeds there that allow them to escape from the Earth. The magnetosphere protects the Earth's atmosphere from the solar wind, but escaping particles can seep out through it, to be eventually swept away by the solar wind. Although, as we have seen, the Earth's magnetic field will decrease, it will provide adequate protection from the solar wind until long into the future. So the question for the future is less about the declining magnetic field than about the number of escaping particles. There is one exception. As we have seen, the direction of the Earth's magnetism flips at intervals. Such reversals take a while to occur. During that time, the Earth is less well protected against the impact of charged particles from space. If so, a magnetic flip in the next few thousand years may require our descendants to plan ahead.

At present, the rate of loss of hydrogen and helium is so small that it can be ignored. The increasing heat from the Sun as it brightens will change this. As we saw in the previous chapter, the temperature of the Earth's surface will eventually reach a point at which a runaway greenhouse effect will occur. Large numbers of water molecules will then enter the atmosphere from the oceans. Some of them will reach the upper atmosphere, where the Sun's radiation will break them down into hydrogen and oxygen. The hydrogen will escape rapidly and be swept away by the solar wind. The oxygen will combine with whatever is handy. Much of it will make its way back to the surface. As the Sun's heat continues to grow, most of the Earth's atmosphere will disappear in this way—either escaping into space or combining with the solid surface. Any remnants that are left will be lost as the Sun evolves from the main sequence. They will be carried off by the huge gusts of material ejected from the Sun as it loses mass. So, though magnetic activity will decrease slowly in both the Earth and the Sun, other events—and especially the increasing brightness of the Sun—will have a much greater impact on the Earth's future.

5. Impact

Meteors and Meteorites

Anyone who watches the sky on a dark, clear night (not so easy with all the light pollution nowadays) will see the occasional flashes of shooting stars. These—more accurately known as *meteors*—are small dust particles which the Earth sweeps up as it moves round the Sun. Like all objects in the inner solar system, they are moving with speeds measured in kilometers per second. When they hit the Earth's atmosphere, the friction produced by such speeds heats the dust particles until they burn up completely. The flash of a shooting star signals its end.

A distinction is usually drawn between a *meteor*, which does not reach the surface of the Earth, and a *meteorite*, which does. To come through the entire atmosphere, a meteorite must be much larger than a dust particle. It has to be able to burn off its outer layers on the way down, and still have enough left to deposit material on the surface. Meteorites are quite capable of depositing many kilograms of rocks, but they occur much more rarely than shooting stars. Typically, only a handful of meteorite falls are recorded each year. Their potential for landing on people's heads is therefore very limited. But the existence of meteorites raises an obvious question. Are there even larger bodies circulating round the inner solar system that could produce more spectacular results if they hit the Earth?

Craters on the Moon

Oddly enough, the best place for answering this question is not the Earth, but the Moon. The Earth's atmosphere acts as a sorting device for the material that it encounters. Small particles are burnt up. Larger rocks can get through, but are slowed down. Big bodies can retain most of their original speed all the way down to the Earth's surface. The Moon has no atmosphere. So, whatever the size of the incoming body, it will retain its original high speed all the way down to the lunar surface. Since no sorting occurs, we can compare the effect of all incoming bodies on the Moon's surface directly. Moreover, we can do this for a long time into the past. Since the Moon has no atmosphere, it has virtually no erosion. Unlike the Earth's very active surface, the Moon's surface is dead. The footprints of the astronauts who landed on the Moon are still likely to be there a million years from now.

If you drop a stone from some height into soft earth, it will create a hole about the same size as the stone. This is the typical result of a low-speed impact. If the stone is moving at the same sort of speed as a meteor, the result is quite different. The impact is explosive, and what results is a crater many times the size of the incoming body. Because the Moon has no atmosphere, everything hits its surface at high speed and produces a crater. A dust particle digs out a tiny crater. A meteorite of the typical size picked up on Earth can excavate a crater meters across. These craters exist on the Moon's surface for a long time. So, by observing the lunar craters and what sizes they have, it is possible to work out the sizes of the objects that the Moon has encountered as it circles the Sun.

There are, of course, complications. One is that when a new body hits the Moon's surface, it wipes out the smaller craters that are already there. But the final result is clear enough. Large craters are rare; small craters are common. Correspondingly, the Moon has encountered a few large objects during its history, but many smaller ones. ("Large" here means objects tens of kilometers across or more.) To some extent, the lunar craters can be sorted into a time sequence. If, for example, a smaller crater appears inside a larger one, then the smaller crater must obviously be the younger. Unfortunately, this does not tell us when either the larger or the smaller crater formed. Was it a billion years ago, or only a hundred million?

The lunar landings, some thirty years ago, helped resolve this problem. The rocks brought back by the astronauts have been dated using the same radioactive techniques that have been applied to terrestrial rocks. The results show that much of the lunar cratering occurred early on. The rate of formation of new craters slowed down greatly between 3.9 and 3.3 billion years ago. By about 2 billion years ago, it had decreased to a fairly low level that seems to have been maintained ever since. Really big objects crossing the Earth–Moon orbit no longer exist: the largest lunar impacts all occurred during the early stage of heavy bombardment. But objects hundreds of meters, or even several kilometers across are still in circulation. The current estimate is that four craters of about 10 kilometers diameter now form on the Moon every ten million years, along with one crater 20 kilometers in diameter.

Impacts on the Earth

Since the Earth and the Moon move together round the Sun, what applies to the Moon can be expected to apply also to the Earth. There are differences—the Earth has more surface area than the Moon and its gravitational pull is bigger. So impacts should create craters on the Earth rather more frequently than on the Moon. The problem with the Earth is that much of its surface is covered with deep ocean. A large body hitting the ocean may well create a tidal wave, but it will not produce a crater. Allowing for this, the estimate is that about three craters of 10 kilometers or more in diameter should be formed on the Earth's land area every million years.

In recent decades, searches have been carried out to try and discover terrestrial impact craters. The difficulty is the changing surface of the Earth. Craters are easily buried, and then can only be found by sophisticated exploration techniques of the sort used by oil companies. The good thing about the craters on Earth is that material can be collected from them. So each crater can be dated fairly accurately. A study of terrestrial craters confirms that their distribution over time is what would be expected from observations of the Moon's surface. But it shows something more: the distribution of ages is not always uniform. Sometimes a series of impacts seems to have occurred relatively close together in time. For example, it has been suggested that about 35 million years ago a number of impacts took place within a period of a million years.

Though the number and scale of the impacts on the Earth's surface are far less now than in the early days of the planet, they may still have unpleasant implications for life on Earth. A tidal wave created by an impact on an ocean can sweep away life on neighboring coastlines. Moreover, big impacts can throw large amounts of material into the atmosphere. For example, an impact on land can throw up a cloud of dust which blankets out the Sun, producing a lengthy period of cold. Although particular species of animals and plants come and go as time passes, there have been occasions when large numbers of species have all disappeared at about the same time. Such a mass disappearance is often referred to as a *biological extinction*. Hardly surprisingly, several attempts have been made to find a correlation between major impacts on the Earth's surface and biological extinctions. One relatively recent extinction happened about 65 million years ago. This was the event that finished off the dinosaurs. It has been claimed that a major factor in this disaster was an impact that occurred just off the Mexican coast, near the Yucatan peninsula. The existence of a crater in the Gulf of Mexico, of the right age and some 180 kilometers in diameter, was confirmed during a search for new oilfields. The explosion needed to produce such a crater would have been equivalent to 100 trillion tons of TNT: quite enough to have global effects. The problem is that major volcanic eruptions were going on in India at the same time. Disentangling the atmospheric effects of the two different events is not easy.

Whether impacts are the sole cause of extinctions is hotly debated. But most people accept them as a contributory factor. One of the questions is how abruptly the environment changes. In the case of an impact, conditions on Earth should change rapidly. For most terrestrial causes, such as volcanic activity, environmental changes should be spread out over a period of time. The impact off the coast of Mexico provides an interesting case study. Beneath the shallow seas in this area, much of the rock consists of carbonates (effectively various metals combined with carbon dioxide). So a large impact here must have put carbon dioxide into the atmosphere along with the water vapor. Plants take in carbon dioxide through pores in their leaves in order to use it for photosynthesis. It has been found that plants in the period immediately after the extinction had fewer pores than plants existing before the extinction. This has been attributed to a rapid increase in the amount of carbon dioxide in the atmosphere. (The plants were trying to control their intake of carbon dioxide in order to keep photosynthesis at a reasonable level.) Such a rapid change fits in well with an impact, but less well with suggested terrestrial causes.

What, then, of the future? Impacts seem to occur at random, so it is not possible to attempt any precise forecasting. Still, the average rate derived above suggests that an impact that will form a crater 10 kilometers in diameter should happen within the next 200,000 years or so. Such an impact will not have a globally disastrous effect—that requires bigger explosions. But it could still have a major local effect. As with bombs, the effect of an impact is not restricted to the area of the actual crater. The explosion makes the surrounding atmosphere expand rapidly, producing a blast wave that spreads over a much larger region. For example, if the incoming body were to hit a city such as New York or London the crater formed would only destroy part of the city. But the blast produced would topple buildings throughout the whole of the city and some way beyond. This blast effect is more important from our human viewpoint than the crater itself.

In 1908, an incoming body is believed to have exploded in Siberia producing a blast wave that blew over people standing as far as 250 kilometers away. No corresponding crater has been positively identified, so it is generally assumed that the incoming body exploded in the atmosphere above the Earth's surface. The explosion injected large quantities of dust into the upper atmosphere, where it spread all round the globe, producing a measurable change in temperature at the surface. So an explosion at one specific site can still have a worldwide influence. What would happen for a similar impact on the ocean can perhaps be guessed from observations of the great volcanic explosion of Krakatau in 1883. Krakatau was a small island in what is now Indonesia. Its explosion caused tens of thousands of deaths, mainly as a result of the tidal wave it created. It too injected material into the upper atmosphere, producing an appreciable lowering of temperature at the surface. (Ground temperatures did not return to normal for five years.)

Impacts on this scale clearly cause global inconvenience, but major effects only locally. The real global problems come from the impact of still larger bodies. Fortunately, these strike the Earth more rarely. Yet "rarely" is a comparative term. On an astronomical timescale, such larger impacts are reasonably frequent—once every 25 million years or so. To put it another way, all the changes that we have looked at in previous chapters suggest that human beings can survive on the Earth for at least another billion years. Over that period of time, there will be something like forty major impacts leading to biological extinctions. Though the blast waves, or tidal waves, from such explosions will destroy life over a large area, it is the atmospheric effects that are the real killer. The materials pumped into the atmosphere are not only great in amount: they also persist for many years. Their specific effects will depend on the place where impact occurs. For example, the minerals on the bed of the Gulf of Mexico include not only carbonates, but also sulfur compounds. If such sulfur is blasted upward, it makes the atmosphere more acid, and so more lethal. Presuming that human beings can survive the altered surface temperature—and possibly changed atmospheric composition—there remains the problem that the food chains on which we rely could be drastically affected. (The destruction of food chains is likely to have been a major factor in past biological extinctions.)

At the really large-scale end of the range, an impact can affect global geology. Studies of the Atlantic and Pacific oceans have shown, for example, that the impact in the Gulf of Mexico caused massive underwater landslides in both oceans. It also produced earthquakes, much larger than anything currently experienced, for many thousands of kilometers in all directions from the impact site. One speculation is that a major impact may even have helped start the breakup of Pangaea. In the long term, as human beings count it, but in the short term on an astronomical timescale, the next major impact will occur. We can always hope that by then our descendants will have devised methods of coping with it. To do that will require a detailed knowledge of the impacting bodies—where they come from and what they are like.

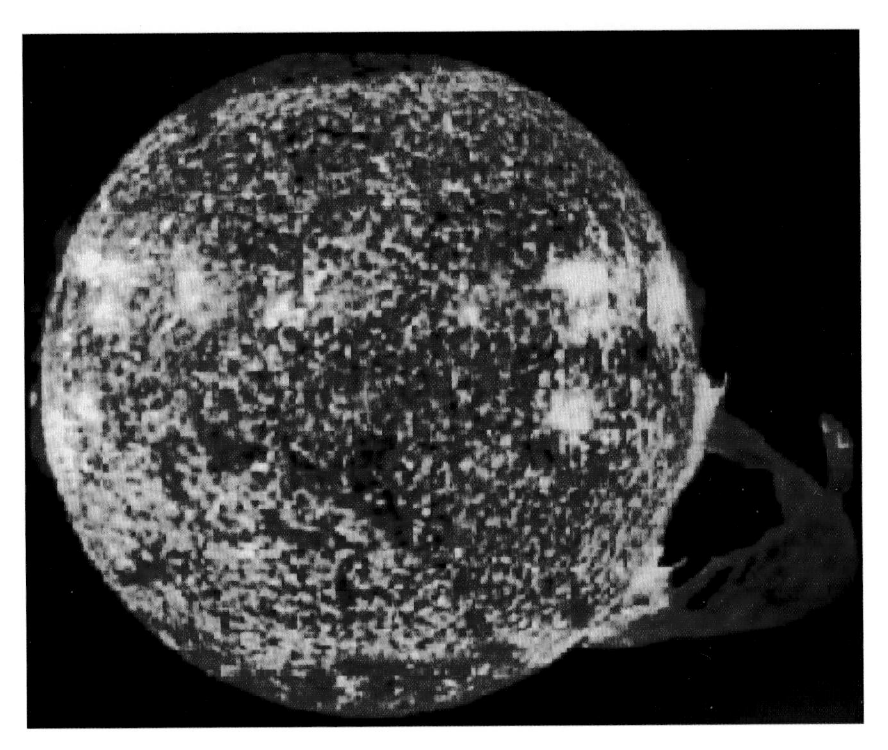

Plate 1 Activity on the Sun

Plate 2 Mount St. Helens erupting

Plate 3 Welcome to Planet Earth

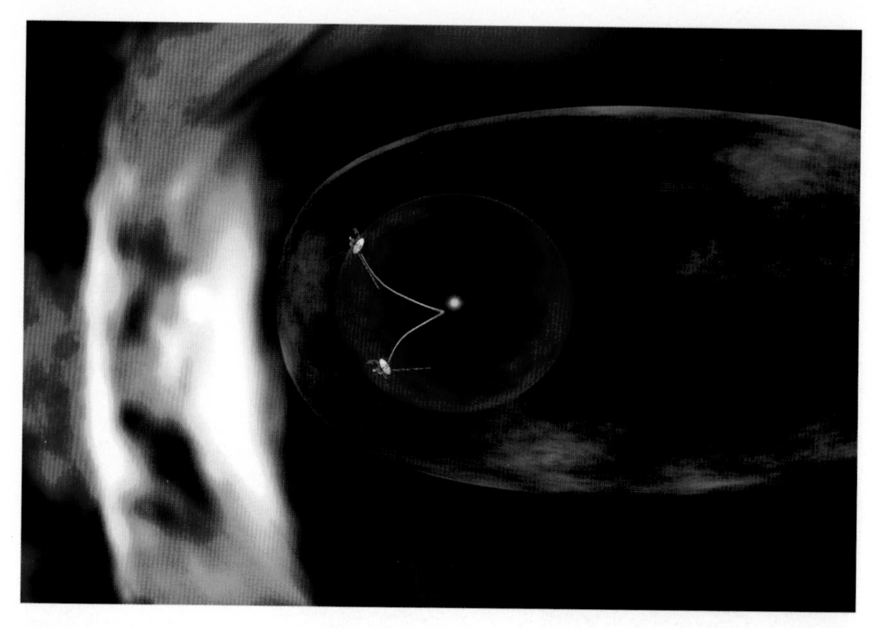

Plate 4 The Voyager satellites reach 90 AU from the Sun

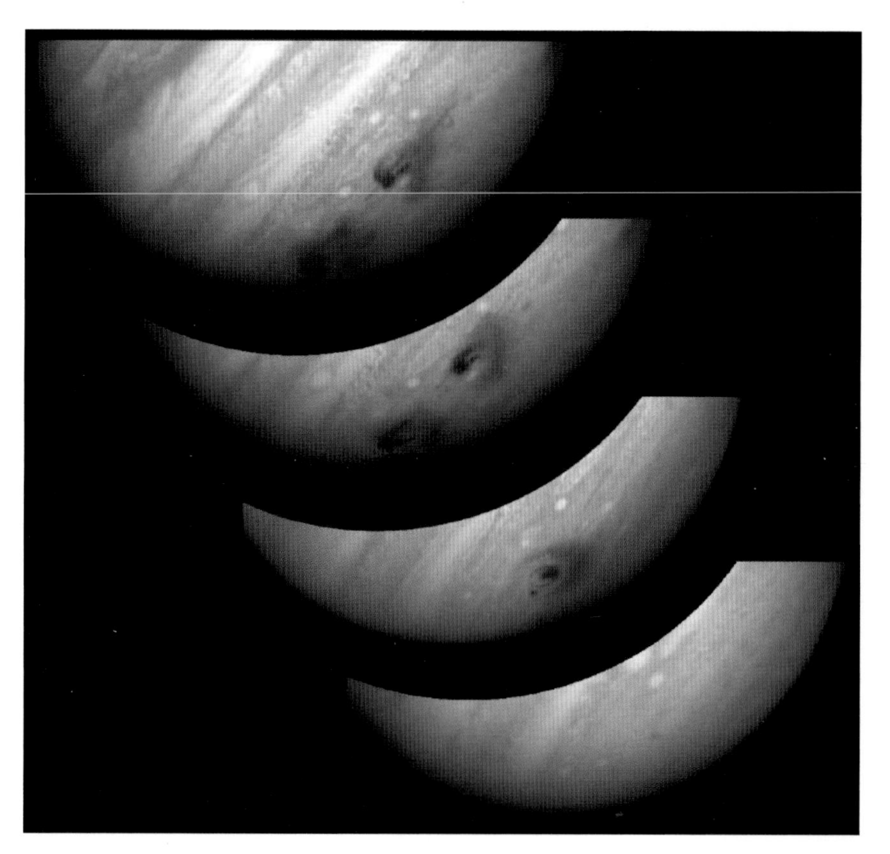

Plate 5 Jupiter Swallows Comet

Plate 6 Solar System Montage

Plate 7 Our Galaxy

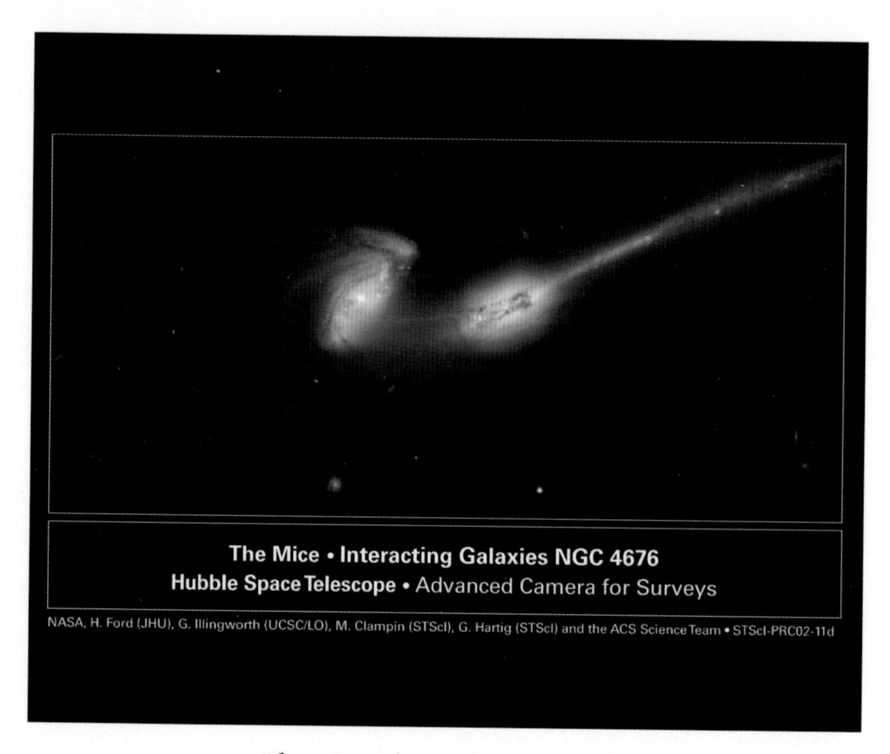

Plate 8 When galaxies collide

Plate 9 Distant Supernova, Dark Energy

The Asteroids

To discuss the smaller objects in the solar system, we first need to picture the layout of the solar system as a whole. The most important object is obviously the Sun. Next in importance come the group of giant planets—so-called because they are much larger than the Earth. (Jupiter, for example, has a diameter over eleven times greater than the Earth's.) There are four giant planets: Jupiter, Saturn, Uranus, and Neptune, listed in order of increasing distance from the Sun. It is customary to measure distances in the solar system in terms of the Earth's average distance from the Sun. This *astronomical unit,* as it is labeled—often abbreviated to "AU"—is equal to some 150 million kilometers. The orbits of the giant planets stretch between 5 AU from the Sun for Jupiter to 30 AU for Neptune. Crammed into the space between the Sun and Jupiter are four much smaller planets. In order from the Sun, they are Mercury, Venus, Earth and Mars. Mercury, the innermost planet, orbits at 0.4 AU from the Sun, and Mars, the outermost of these small planets, at 1.5 AU. For the most part, the planets orbit the Sun in round about the same plane. In other words, drawing the orbits on a flat piece of paper is not too bad a representation of what is happening. This plane is called the *ecliptic.* Not only do the planets not stray too far above or below the ecliptic: they also all move round the Sun in the same direction.

A look at the spacing of the planets suggests that there is a surprising gap between the inner, or "terrestrial" planets, and the outer, or "Jovian" planets. Two hundred years ago, it was discovered that this boundary region between the two types of planet was occupied by a number of small bodies, subsequently called the *asteroids.* The number of asteroids known has increased rapidly with time. Indeed, the main problem nowadays is not discovering them, but keeping track of them. Most of them are small. Only a handful are more than a hundred kilometers across, but hundreds exceed ten kilometers. There are very many asteroids less than a kilometer across. They are hard to detect directly because, given their distance from the Sun, they reflect little light to us. Our main interest is in their potential for hitting other bodies in the solar system in the future. More especially, we are interested in bodies that might hit the Earth.

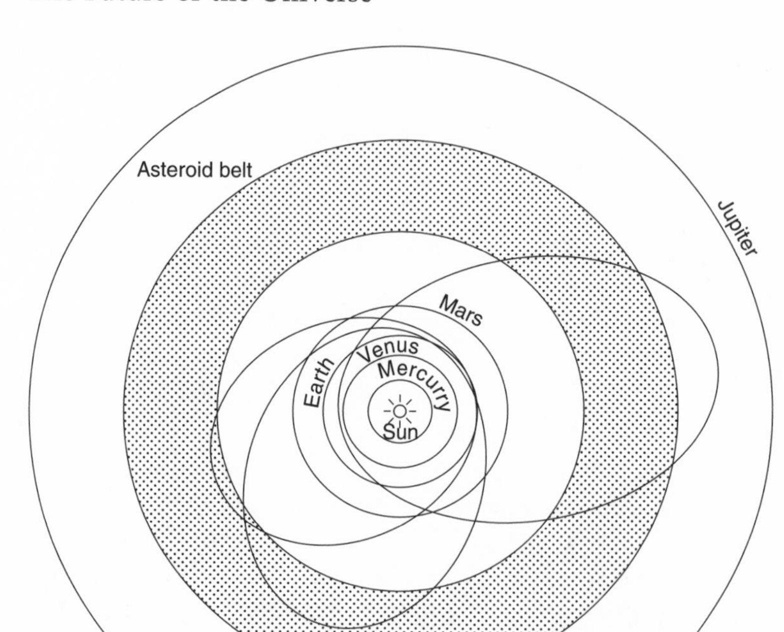

5.1 The position of the asteroid belt between Jupiter and the inner planets. Most asteroids have fairly circular orbits, but material in the asteroid belt can sometimes be deviated inwards by Jupiter's gravitation. It may then collide with one of the inner planets. The orbits shown are for three meteorites that have actually been observed hitting the Earth.

The asteroids follow a variety of paths round the Sun. They spread out over a wide belt with most, but not all, asteroids orbiting at distances between 2.2 and 3.3 AU from the Sun. Jupiter is easily the most massive planet in the solar system and it is quite close to the asteroid belt. It has a strong gravitational effect on the nearby asteroids, and is a major factor in the irregular distribution of the asteroids in the belt. In some cases, Jupiter's influence makes an asteroid's motion round the Sun unstable, and pulls it into a new orbit. Some of these new orbits may intersect the paths of the terrestrial planets, not least the orbit of the Earth. Since most of the bodies in the asteroid belt are small, so is most of the material that crosses the Earth's orbit. This is why meteorites are not usually very massive. Even so, the bigger ones can do some damage. The Barringer crater in Arizona is regarded as a typical example of a meteorite crater. It was created by the impact of a meteorite some tens of meters across, which struck the Earth some 50,000 years ago. The crater which it gouged out is 1.2 kilometers in diameter and 200 meters deep. Such an impact, though leaving an impressive scar, has little effect on anything outside the immediate neighborhood. But it has been estimated that many much larger bodies than the Barringer meteorite are currently circulating through the inner solar system. There are probably over a thousand with diameters of a kilometer or more that can intersect the Earth's orbit. It is amongst this group that we must look for potential impacts leading to future biological extinctions.

Forecasting Future Impacts

The problem in trying to forecast impacts on the Earth (and on the other planets) is that the situation is continual changing. The asteroidal material in the inner solar system is affected by the various gravitational pulls of the terrestrial planets, as well as continuing interactions with Jupiter. Asteroids with paths that currently cross harmlessly above or below the Earth's orbit may, in time, be diverted into paths that are actually on a collision course with the Earth. For example, Eros is an asteroid with a well-established orbit. At present, its orbit does not intersect the Earth's. According to computer calculations, there is a 10 percent chance that, some millions of years in the future, Eros will be diverted into a new path which does cross the Earth's orbit. With a diameter of 35 kilometers, Eros is estimated to be three times larger than the asteroid that caused the biological extinctions 65 million years ago. The effect of an impact by Eros would correspondingly be very much larger.

Over time, some of the asteroidal material in the inner solar system will fall into the Sun; some will be ejected into the outer solar system; some will collide with the inner planets. The timescale on which the pieces disappear depends on their particular orbits and on the gravitational pulls that they encounter. Many only survive for 100,000 years, whereas others may remain in circulation for much longer. Their disappearance presents us with an immediate question. Counts of the craters on the Moon suggest that impacts in the inner solar system have happened at a fairly regular rate over the past 2 billion years. Yet the bodies available in the inner solar system to produce the impacts mostly have quite short lifetimes. How do we fit these two facts together?

Gravitational interaction, mainly with Jupiter, ensures that material from the asteroid belt is continually being diverted in toward the Sun. Though the interaction may vary a bit with time, it remains sufficiently constant to ensure that the amount of material moved inward does not change much over long periods. If material is continually being removed from the asteroid belt (and Jupiter throws some outward, as well as inward), the quantity of material remaining must be decreasing. Why is this reduction not reflected in the record left by the impact craters? The answer is that the asteroids circulating in the asteroid belt have a fair chance over long periods of time of hitting each other. If a small piece hits a large piece, the latter will develop a nice new crater. If the two bodies are more equal in size, the collision may well break them up altogether. Instead of the initial two bodies, there will now be thousands of smaller bodies. Some of the resulting fragments will be thrown into orbits which will ultimately loop into the inner solar system. What is happening is that larger objects in the asteroid belt are being degraded into numerous smaller fragments. This increase helps keep the number of fragments entering the inner solar system constant.

The range of sizes of the fragments produced by such collisions actually fits in very well with the size distribution indicated by the lunar craters. In fact, the same range of sizes can be reproduced on a small scale by simply hitting a piece of rock with a hammer. The result is always a few large pieces, a fair number of more moderate size, and many small ones. It is true that the mass of material in the asteroid belt is decreasing, but the number of bodies in circulation is not. So far as the inner solar system is concerned, lost asteroidal material is continually being replaced by new fragments from the asteroid belt. Consequently, the cratering rate remains fairly steady.

Working out the rate at which impacts have occurred is not easy to do accurately. But analysis of the number of craters on the surfaces of the Moon and the Earth suggests that asteroidal material is unlikely to be the only source for crater formation. More impacts are recorded than would be expected from the number of asteroidal fragments currently circulating through the inner solar system. Fortunately, it is easy to identify other bodies out there that can also create craters.

Comets

A distinction has traditionally been drawn between two types of small body found in the solar system—asteroids and comets. This has been based on two things: the paths they follow round the Sun and their differing appearances. Most asteroids lie in a belt this side of Jupiter, and typically follow fairly circular paths round the Sun. Comets come in from beyond Neptune and move round the Sun in elongated, elliptical paths.

Asteroids are made of rock. Comets seem to be made mostly of volatile material, such as water or ammonia. Beyond Neptune, the Sun's heat is sufficiently small for these substances to remain frozen. As the comets move in toward the Sun, the temperature rises, and the volatile material begins to evaporate. The gas produced is broken down into electrically charged particles by the Sun's radiation. These particles are then swept backward by the solar wind, forming a lengthy tail—the characteristic feature of a comet. The tail also contains dust particles left behind by the comet, and these are the commonest source of shooting stars. In fact, comets contain rock as well as volatile material. The mixture can create impact craters on a planetary surface just as readily as asteroidal fragments can. The question is whether comets inject material regularly into the inner solar system, as asteroids do. To answer that requires knowledge of where comets come from.

In the last decade or so it has become apparent that the division between asteroids and comets is not absolute. Moreover, a large number of small bodies have been found orbiting the Sun at distances well beyond Neptune. The region they occupy has been labeled the "Edgeworth–Kuiper belt," after the two people—one on the east of the Atlantic Ocean and the other on the west—who first suggested its existence. (Somewhat unfairly for Edgeworth, this is often now abbreviated to the "Kuiper belt.") The inner edge of this belt starts at about Neptune's distance from the Sun (30 AU). Where its outer edge lies is anybody's guess. Some suggestions put it as far out as 1,000 AU. Fortunately, it is the inner edge of the belt that concerns us here: the exact extent of the belt is not significant.

Like the bodies in the asteroid belt, those in the Kuiper belt have orbits that cluster round the ecliptic. In other words, they circulate round the Sun in much the same plane as the planets. As with the asteroid belt, this leaves them open to interactions with the outer planets—in this case, especially Neptune. From time to time, one of the bodies in the belt is deviated in such a way that it dips in toward the inner planets, where it loses its volatiles, so that we see it as a comet. Technically, it is known as a short-period comet, meaning that it returns to the inner solar system after a period measured usually in tens of years. As would be expected, these Kuiper belt comets circulate close to the ecliptic plane. Yet a few short-period comets orbit the Sun at a considerable angle to the ecliptic. A famous example is Halley's comet, which takes 76 years to orbit the Sun. It not only follows a path that goes a long way out of the ecliptic; it is also moving round the Sun in the opposite direction to the way the planets move. It is difficult to see how such comets could come from the Kuiper belt.

The short-period comets are so-called in order to distinguish them, unsurprisingly, from the "long-period comets." "Long-period" here may mean that the comet takes a million years to go once round the Sun. These comets are believed to come from yet another group of small bodies, in this case lying at distances from a few hundred AU to perhaps 100,000 AU from the Sun. Unlike the objects in the Kuiper belt, these more distant objects follow orbits that can make any angle to the ecliptic. They lie in a shell that surrounds the Sun in all directions. (It is known as the "Oort cloud," after the Dutch astronomer who first looked at the problem of long-period comets in detail.) At distances like this, even the giant planets are too far away to have a significant gravitational effect. But events outside the solar system—which will be discussed in a later chapter—can occasionally provide a stimulus that sends one or more of these distant bodies tumbling in toward the Sun.

The great majority of such comets enter the inner solar system only briefly, spending most of their time far out beyond the planets. For a few, it is different. As they pass through the central parts of the solar system, they chance to stray close to one of the giant planets, and are captured by its gravitational pull into short-period orbits. Halley's comet is thought to be one of these. The same disturbance that sends comets inward also sends them outward. Many comets in the outer parts of the Oort cloud must have been lost to space. Presuming that other stars have similar clouds, there must be a considerable number of comets wandering about between the stars. Some may encounter the solar system, offering us the chance of seeing interstellar comets. Unfortunately, the probability of seeing one is quite small. At best, one visible comet every 150 years or so may be interstellar in origin.

We now have two additional sources, besides the asteroids, that can supply small bodies to the inner parts of the solar system—the Kuiper belt and the Oort cloud. (There are probably other places where small bodies can be found in the solar system. They can be ignored here, since objects from these parts rarely seem to wander into the inner solar system.) The question is how frequently comets from these two sources can produce craters on the Earth and the other terrestrial planets. The answer seems to be that the impact rate fluctuates both with time and with the size of the crater. At present, comets are less likely to hit the Earth than asteroidal fragments. But the mechanisms that force comets inward, especially from the Oort cloud, may lead to several appearing in sequence. At such times, the probability of cometary impacts occurring obviously goes up.

Comets become more important as the scale of the impact goes up. They are more frequently the source of the larger craters. It has been estimated that perhaps half of all the craters on Earth with a diameter of more than 50 kilometers are the result of cometary collisions. This rises to 80 percent for craters with diameters of more than 100 kilometers. Because the bigger craters are usually produced by cometary impacts, major biological extinctions are believed to be due mainly to comets. Since they contain large quantities of volatile material, comets produce more complicated atmospheric effects than asteroidal fragments. They are also more likely to explode in the atmosphere, which is an effective way of creating blast waves. (It has been speculated that the explosion over Siberia in 1908 was caused by a small cometary fragment.)

We can raise the same question here for comets as we did for asteroids. As time passes, the number of comets must diminish. Does this raise any problems regarding provision of new comets for the future? Collisions leading to fragmentation have certainly occurred over past history in the Kuiper belt. Moreover, the belt as a whole currently contains more material than the asteroid belt. So it is a fair bet that short-period comets will continue to appear while there is anyone around on Earth to notice them. For long-period comets, the position is even clearer. To produce comets at the rate we see them, there must be an enormous number of them in the Oort cloud. Estimates suggest a figure of well over a trillion (a million million). There is no way that we can run through this number in the next few billion years, unless something extraordinary occurs.

Of all the astronomical events that may cause discomfort to our successors on the Earth, impacts by small bodies rank high. Bodies 100 meters in diameter may hit the Earth once every

5.2 The Chicxulub crater in Mexico. It gives some idea of the size of the impact needed in order to produce world-wide consequences.

thousand years on average. These can have significant local effects. Really major impacts occur perhaps once every 30 million years, so we hardly need to feel that the sky will fall on our heads immediately. But sooner or later our successors will be called on to cope. Will they be able to do anything about it? The obvious first step is to try and pin down how many objects currently have a chance of hitting the Earth. Such a survey is already under way. For asteroidal fragments, the search is not too difficult. With good enough instrumentation, they can be observed and their paths calculated. Comets from the Kuiper belt are a bit more difficult, because they are typically too faint to be followed over parts of their orbits. The really difficult ones are comets from the Oort cloud. The likelihood of them hitting the Earth can only be calculated after they have first become visible on their inward journey. So long as a reasonably close eye is kept on the heavens, it should be possible to pick them up by about a year before they actually hit the Earth. The overall problem is that probably no more than 10 percent of the sizable objects that can cross the Earth's orbit are currently known.

Supposing that a potential Earth-collider can be identified, there still remains the question of what to do about it. The currently favored option is to dispatch a space probe to the oncoming object, and use rockets on the probe to nudge the fragment into a safer orbit. How this might be done depends on the physical nature of the object. Some are solid objects that can be pushed or pulled. Others are weakly bound fragments that would easily fall to pieces. In some cases, the only route might be to blow the object up entirely, so that should any fragments hit the Earth they would be too small to produce big craters. These are the thoughts in circulation today. We can hope that our descendants will be a good deal cleverer than we are, and will come up with better solutions.

6. The Solar System

The layout and visual appearance of the planets in the solar system make it obvious that they fall into two groups—the inner, terrestrial (Earth-like) planets and the outer, Jovian (Jupiter-like) planets. The two groups not only have different properties; they are likely to have rather different futures. From an Earth-based perspective, the inner planets—Mercury, Venus, Earth and Mars—are more immediately interesting. We can lump in with them the Earth's Moon and the two small satellites that circle Mars. As it happens, the Earth and the Moon between them nicely bracket the properties of the terrestrial planets. On the one hand, the Earth is active, with things happening in its interior, on its surface, and in its atmosphere all the time. On the other hand, the Moon is dead, with almost no atmosphere and with few changes occurring in its interior or on its surface. We can characterize the other terrestrial planets in terms of where they lie on this scale of activity: closer to the Earth, or closer to the Moon?

The Moon and Mercury

The previous chapters have shown that the Earth's future will be as complicated as its past. The Moon's future, on the contrary, is likely to be a good deal simpler than its past. As we have seen, one thing that will change slowly is the Moon's orbit round the Earth. There will also be occasional changes to its surface, since additional comets and asteroidal fragments will impact on it as the years pass. But these impacts will make only minor differences to its overall appearance. In the early days, flows of molten rock formed the dark patches that we see on the Moon. The conditions no longer exist for features like these to recur in the Moon's old age, so our naked-eye view of the Moon will change little. Though the Moon is still cooling down, future temperature changes inside will not lead to any major alteration in its structure. It has little by way of a conducting core, and it only rotates once a month (which is why we always see the same face). So there will be no changing magnetic field.

Mercury, the innermost planet, has much in common with the Moon. It is somewhat larger and appreciably denser, but its cratered surface looks generally similar. As with the Moon, the vast majority of the craters were formed long ago. Mercury is denser than the Moon because it contains much more iron in its interior. This is probably why it has a small magnetic field, presumably produced originally by dynamo action. If so, the activity probably stopped some time ago, for Mercury has cooled, and most of its iron core should now be solid. In addition, the planet is spinning only slowly—about once every three months in Earth terms. So the remaining field is a bit like that of a bar magnet, and it will disappear slowly as time passes.

Apart from occasional impacts, the main excitement for Mercury is likely to be in the distant future. At high noon on Mercury, the surface temperature currently reaches some 500 °C. This is already above the melting point of such metals as lead and tin. As the Sun brightens with time, the maximum temperature on Mercury's surface will rise further, past the melting points of such metals as aluminum or magnesium, and will continue to rise until the planet is consumed by the expanding Sun. Yet there is a faint possibility that the planet will no longer be waiting there to be swallowed. Mercury follows an elongated orbit round the Sun. The actual shape of the orbit changes with time due to interaction between Mercury and the gravitational pulls of the other planets. Since Venus is the nearest planet to Mercury, it has a particularly strong effect. Over the years, it is possible that Mercury's orbit will move further out from the Sun, and so come quite close to Venus. At this point, Venus can affect Mercury so greatly that it runs away from the Sun altogether, and is lost to the solar system. It has been calculated that this could happen within the next 3–4 billion years: that is, before the Sun becomes a red giant.

Venus

The next planet out from the Sun is Venus. If Mercury has much in common with the Moon, Venus has much in common with the Earth. Its diameter is some 5 percent less, but the internal structure is believed to be very similar. Like the Earth, the planet has a solid mantle and a liquid core (though not a solid inner core). What it lacks is a magnetic field. The reason for this may be its very sluggish spin rate. One day on Venus is equivalent to about two-thirds of a year on Earth. But there may also be differences in the core: too little is known about it to be sure. Yet, oddly enough, the overall similarity between the interiors of Venus and the Earth has not led to identical activity at the surface.

At first glance, the surface of Venus has features in common with the Earth's surface. There are volcanoes, along with a number of impact craters. Most of the surface is low-lying, like the ocean basins on Earth, but some is more elevated, like the terrestrial continents. The crucial difference is that Venus shows no signs of the division into plates that is so characteristic of the Earth's surface. In the Earth's mantle, the heat from the core is transferred efficiently to the surface by loops of convective material which have up-and-down motions in the interior and sideways motions at the surface. On Venus, all the movement is essentially up and down. Consequently, while terrestrial volcanoes tend to cluster along plate boundaries, volcanoes on Venus are distributed much more at random. They are not linked with regions where crust is descending into the mantle. We saw that the Earth carried some of its heat upward via hot plumes that rose from the edge of the core out toward the crust. This kind of activity seems common on Venus, where it produces all the major volcanoes.

Impact craters appear all over the surface of Venus, but in much smaller numbers than on Mercury or the Moon. Indeed, the number per unit area of the surface is much like the number of terrestrial impact craters. Craters on Earth are relatively scarce because much of the Earth's surface is being constantly reworked by plate motions. On Venus, reworking of the surface is apparently due to big plumes—superplumes—which bring up molten material from the mantle and spread it out across large areas. Such resurfacing events wipe out previous impact craters, starting the clock back at zero. From the number of craters currently visible, it can be estimated that the last global change of this kind must have occurred several hundred million years ago, though there has been smaller-scale volcanic activity since then. Presuming that the outpourings happen every time too much heat gets dammed up in the interior, it can be estimated that the next global resurfacing will occur a few hundred million years from now. This means that Venus can expect four or five such events over the next 1–2 billion years, each one making a major change to the Venusian landscape.

The real question, of course, is why the surface of Venus does not have mobile plates like the Earth. The culprit is believed to be the near absence of water on Venus. Water on Earth acts as a lubricant to assist up-and-down motions near the terrestrial surface. The surface of Venus, without such water, seizes up, forcing the heat to escape in a different way. Indeed, the low amount of water on Venus can be used to explain several of its peculiarities. On the Earth, much of the carbon dioxide in circulation is found as solid carbonates, whose formation requires the presence of water. On Venus, without liquid water, the carbon dioxide remains in the atmosphere in massive amounts. This produces a large greenhouse effect, leading to a surface temperature for Venus of some 450 °C. Consequently, even if liquid water were to appear at the surface of Venus, it would boil away immediately.

Venus does not experience seasons in the same way as the Earth. Its spin axis is nearly perpendicular to its path round the Sun. Moreover, this angle is unlikely to change much with time. Venus, unlike the Earth, does not have a moon to stabilize its spin axis, but tidal interaction with the Sun provides a sufficient stabilizing force. So there is no such thing as Milankovitch variations for Venus. In any case, the slow rotation and thick atmosphere of Venus combine to redistribute heat efficiently round the planet. Variations in temperature between the equator and the poles are not important as they are on Earth. The wind pattern is quite different from the Earth's, and wind speeds are lower.

Even though Venus cannot have water on its surface, we would expect, by comparison with the amount present on its sister planet Earth, that there should at least be a lot of water vapor in its atmosphere. In fact, there is very little. Even the clouds on Venus consist of sulfuric acid, rather than the water droplet clouds we have on Earth. This is not to say that Venus never had any water. As with the Earth, molecules that drift into the upper atmosphere of Venus are exposed to ultraviolet light from the Sun, which breaks them down. So any water there is broken down into hydrogen and oxygen. Venus has no protecting magnetosphere, and hydrogen escapes into space more readily than on Earth. Meanwhile, the oxygen left behind recombines with whatever else is present—ultimately the planet's surface. The overall result is the removal of water from the atmosphere.

With this background, we can ask what long-term changes in climate Venus might experience in the future. Current volcanic emission of gases seems too small to change the nature of the atmosphere significantly. A global resurfacing event is something else. Such an event will be accompanied by a major release of gases from the mantle, especially water vapor and sulfur dioxide. These react to form massive new sulfuric acid clouds, which shield the surface, causing its temperature to drop below the present value. This acid then reacts slowly with the surface over a period of 200 million years or so, thinning the clouds in the process. The surplus water vapor in the atmosphere now enhances the greenhouse effect, making the surface temperature rise to a value higher than that existing today. After a further 200 million years, it may reach a peak of 600 °C. Subsequently, the slow loss of hydrogen from the upper atmosphere reduces the amount of water vapor, and gradually returns the atmosphere and the surface temperature to something like their present states.

This climatic cycle can be expected to recur several times over the next 2 billion years. The increasing brightness of the Sun will affect the details of the process. In particular, the rise in temperature should make the loss of hydrogen easier, gradually speeding up the cycle. But there may be another way of changing the atmosphere of Venus. Comets contain large amounts of water in the form of ice. A moderately large comet impacting on the surface of Venus could double the amount of water in the planet's atmosphere. Such impacts may be as common as global eruptions on Venus. They differ in occurring at random, whereas the eruptions occur at spaced intervals. The additional water acts to increase the greenhouse effect, and so the surface temperature. Comets can therefore be expected to supplement the effect of internal activity in the future. Ultimately, as the Sun continues to brighten, the atmosphere of Venus will be blown away. The planet itself will then be gulped up as the Sun expands into its red-giant phase.

This discussion of Venus has underlined the importance of water in the evolution of the terrestrial planets. It is customary to talk of the *habitable zone* round a star, defined basically as the region round the star where liquid water can survive on a planetary surface. (As the name "habitable zone" tells us, liquid water not only affects the geology of a planet, it is also essential for life.) In the present solar system, the calculated inner limit for the zone lies between the orbits of the Earth and Venus and the outer limit lies between the Earth and Mars. The unsurprising result is that only the Earth has extended areas of liquid water on its surface, and is therefore easy for life forms to inhabit. The limits to the zone obviously depend on the brightness of the Sun. In future, the limits will move outward. This will simply make Venus even hotter than it is today. It will also ultimately trigger a runaway greenhouse effect on Earth. But what will it do to the more distant Mars?

Mars

Mars is smaller than the Earth or Venus: not much more than half their size. Consequently, it shares some properties with them, but it also shares some with the smaller Mercury. Mars is thought to have a core, mantle, and crust, like Venus, but it has no significant magnetic field. This is surprising, since the planet is spinning quite rapidly: a day on Mars is only slightly longer than a day on Earth. A liquid core in a planet spinning with this speed should certainly allow a dynamo to develop. But the Martian core is estimated to be quite small, and it will have cooled more rapidly than the cores of Venus or the Earth. One possibility is that the core has now solidified to an extent that prevents a fluid dynamo from operating.

If Mars has lost heat from its interior at a more rapid rate than the Earth, then it would be expected to have a lower level of internal activity. This conclusion is supported by observations of the Martian surface. Much of it is old—over three-quarters probably formed more than 2.5 billion years ago. This can be deduced from counts of craters on the planet's surface. The northern parts of Mars are younger on average. Here there are signs of volcanic activity and possibly of superplume development. The youngest volcanic regions on the Martian surface may even have produced flows of lava in the last few million years, but the areas involved are small. Even in northerly latitudes, most of the Martian surface is much older. The picture we have is in some ways like Venus. Vertical motions have dominated in the Martian mantle, and heat loss has mainly been by conduction through a gradually thickening lithosphere. The difference is that heat flow from the interior of Venus is still large enough to create major gushes of molten rock across the surface at intervals. On Mars, the smaller flow of heat means that volcanic activity is on a smaller scale. In the future, impacting bodies are more likely than internal activity to produce changes to the surface.

Mars has only a thin atmosphere, as might be expected from the smallness of the gravitational pull at its surface. What there is consists mainly of carbon dioxide, along with a small amount of water vapor. Since Mars is spinning at much the same rate as the Earth, it might be expected to have similar wind systems. Basically, this is true, but interactions with the surface lead to differences. Temperatures on Mars can fall very low (the average daily temperature is about −50 °C) owing to its thin atmosphere and distance from the Sun. The consequence is that as much as a third of the atmosphere deposits itself on the surface round the poles in the winter. This removal of a large proportion of its atmosphere with the seasons is unique to Mars.

At present, the angle that the spin axis of Mars makes with its orbit round the Sun is very similar to the inclination of the Earth's axis. This means that Mars has seasons like our own. The difference is that Mars moves in an appreciably elongated orbit, so that Martian winters and summers are not the same length in its northern and southern hemispheres. In addition, the tilt of the axis can change much more for Mars than for us. We have a massive Moon to stabilize the position. Mars does not, and is, moreover, greatly affected by the gravitational influence of its neighbor, Jupiter. As a result, its angle of tilt can vary from 15° to 35° over a period of 100,000 years. Such a change leads to major differences in the seasons and consequently in the condensation of gases at the poles. This seems to be the cause of the layered appearance of the polar caps on Mars, and will obviously operate in the future as in the past. Over longer periods of time—tens of millions of years—the changes of angle can be even larger, so that the polar caps can either disappear altogether or spread much more widely.

One of the intriguing features on the Martian surface is what seem to be dried river beds. Since there is little water in the planet's atmosphere, and any on its surface is frozen, where do these features come from? It is generally believed that Mars had a thicker atmosphere earlier in its history. The shield that this provided would have allowed extended areas of liquid water to form on the surface. (The recent results from the American landings on Mars seem to support the existence of such "lakes.") Since those days, Mars has lost a considerable amount of atmosphere. The planet has a relatively weak gravitational hold on the molecules in its upper atmosphere, and, in the absence of a magnetic field, the passing solar wind can readily sweep them away. Some water seems to have remained behind, but as underground deposits below the surface of Mars. A cometary impact on the surface of Mars would add considerably to the present thin atmosphere. It would also heat up and release water from the underground reservoirs, so that for a short period of time rivers might flow again over the surface. But then much of the new atmosphere would be lost to space, and the surface would revert to its present state. Still, such events should change the appearance of Mars on a number of occasions over the next few billion years.

The major long-term influence—as with the other terrestrial planets—will be the growing brightness of the Sun. In principle, this should make Mars a pleasanter planet to live on than it is at present, since its surface temperature will move toward that of the present-day Earth. Such underground water as is left by then will seep to the surface. The problem is that, because the heating process is slow, water reaching the surface will be lost to space as rapidly as it appears. So Mars will not develop an Earth-like appearance. Indeed, not only the water on Mars, but its atmosphere as a whole, will gradually disappear. Ultimately, as the Sun's temperature continues to grow, the Martian surface will come to look more like the present-day surface of Mercury. If the Earth fails to survive the expansion of the Sun to the red-giant stage, Mars will remain as the sole representative of the terrestrial planets still circling the Sun. But its orbit—now more distant from the Sun—will not remain stable for ever. Even if Mars survives the outflow of material from the extended Sun, within the following billion years or so it is likely to wander away into space.

Mars has two satellites, but they could hardly be more different from the Earth's massive Moon. Both are small lumps of rock only a few kilometers across. Since Mars orbits next door to the asteroid belt, it is generally supposed that both moons are actually asteroids that Mars has managed to capture. The inner moon, Phobos, is very close to Mars and is actually in an unstable orbit. Tidal interaction with Mars is forcing it to spiral down: it is calculated that it will hit the surface of Mars in about 50 million years from now. In fact, it is more likely to break up as it gets close to the planet, providing the planet with a temporary ring. The fragments in this ring will later rain down into the Martian atmosphere. Because Mars has a fairly feeble gravitational pull, it does not capture asteroids easily. Still, over the next few billion years, it should be capable of capturing (and losing) occasional lumps from the asteroid belt.

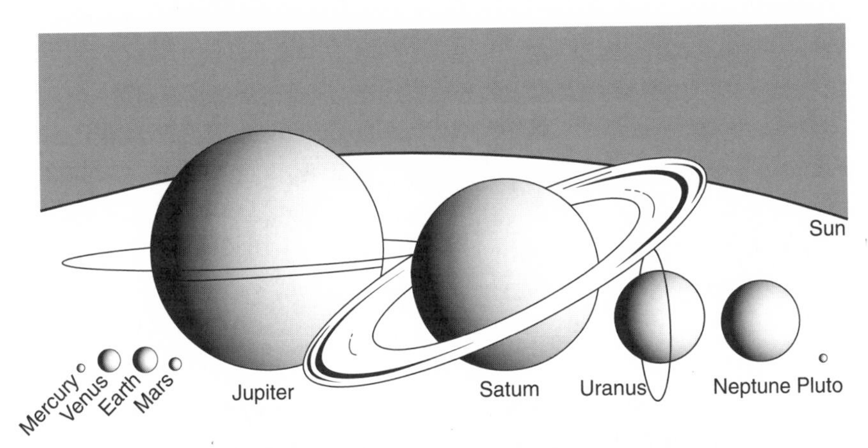

6.1 The relative sizes of the planets and the Sun. The planets are shown from left to right in order of their distance from the Sun. The difference between the inner terrestrial planets and the outer Jovian planets is obvious.

The Giant Planets and Their Satellites

The giant planets present an entirely different picture. We can take Jupiter, the best-studied of these planets, as an example. It is over 300 times more massive than the Earth, and in many ways has more in common with the Sun than with the Earth. Like the Sun, but unlike the Earth, it has no clearly defined solid surface. Again like the Sun, it is made up primarily of hydrogen and helium. Moreover, it is producing an appreciable amount of heat in its interior: almost as much as it receives from the Sun. It is spinning rapidly (about once every 10 Earth hours), and has a large magnetic field. In this case, the magnetism is connected with its hydrogen content. When highly compressed at relatively low temperatures—as in the interior of Jupiter—hydrogen becomes electrically conducting. This, along with convection currents in the interior and the rapid spin of the planet, then produces a dynamo effect, as in the Earth or Sun.

The convection currents are created by the internal heat as it tries to escape from Jupiter. The heat comes mainly from the gradual contraction of Jupiter: the same source as the Sun had for energy in its early days. Jupiter, having much less gravitational pull than the Sun, contracts more slowly. Another energy source—currently more important for Saturn than for Jupiter—is the separation out of hydrogen and helium. Helium is a heavier element than hydrogen. Over long periods of time, it will tend to settle down below the hydrogen. This is another form of gravitational contraction, and so also produces heat. Looking to the future, Jupiter should continue producing internal heat throughout the Sun's career on the main sequence, but at a decreasing rate. At present, the Sun's contribution to heating the atmosphere of Jupiter is only a little bit larger than the planet's own contribution. As time passes the difference will increase, with the Sun providing more and more of the heat. This will produce some changes in the appearance of the Jovian atmosphere (which is the only part we can see). But the basic pattern of bands of cloud—numerous on Jupiter because of its rapid rotation—will continue.

Saturn has considerable similarity to Jupiter: Neptune less so, and Uranus even less. For both Saturn and Neptune, the amount of heat given out by the planet is of the same order as the amount received from the Sun. (Though it has to be remembered that these are more distant planets, and so receive much less solar energy than Jupiter.) No internal heat has yet been measured from Uranus. As with the terrestrial planets, the giant planets must eventually lose their internal heat. Because of the greater size of these planets, the timescale for cooling down is longer than for the terrestrial planets. Indeed, the planets will experience the additional heating from the Sun as it brightens before their own internal heat has dwindled away altogether. The change in the balance between internal and external heating will alter how they look to an observer, since the Sun's heat will particularly affect how their atmospheres circulate.

The giant planets have several advantages over the terrestrial planets in their struggle to survive the Sun's long-term evolution. They are further away, they are more massive, and they have large protective magnetospheres. This means that they are better protected from increases both in the Sun's heat and in the increasing amount of material that it throws off into space. As a result, the giant planets are likely to survive the expansion of the Sun, though not without considerable loss of material. The problem is to know whether they will continue to follow stable orbits. Currently, the gravitational interactions between the planets are such that they follow fairly well-defined paths round the Sun. As the Sun loses mass, the planets will circle further and further away. Will they continue to interact in such a way as to make their new orbits stable? The answer depends on a variety of factors—for example, on how rapidly the Sun loses mass. But it seems that one or two of the giant planets, and perhaps all four, should survive until the Sun becomes a white dwarf. This is not quite the end of the story, for no orbit is entirely stable over time. After a period of 10 billion years or more, even the surviving giant planets will wander off into space (presuming that they do not collide with each other or the Sun).

The giant planets have one more feature that distinguishes them from the terrestrial planets – their attendant trains of satellites. Those of their satellites that lie close to their parent planet, are often comparable in size with our own Moon, and move in fairly circular orbits round the planet's equator. Further away, the satellites are small and have a variety of orbits, often elongated and inclined at a considerable angle to the equator. Satellites with regular orbits are thought to have been born with the planet. Those with more unusual orbits have mostly been captured more recently. (Since the giant planets lie between the asteroid belt and the Kuiper belt, they have plenty of opportunities for capturing additional small bodies.)

Some of the irregular satellites bunch together into rather similar orbits. The deduction is that some of the original bodies that were captured broke into pieces. Fragmentation has also occurred in the rings of small particles that surround all the giant planets (though Saturn's are by far the most spectacular). Such ring systems are not stable in the long term owing to collisions between the particles. It has been estimated, for example, that Saturn's main rings can only last for another 100 million years or so, and more fragile rings may disappear in a few thousand years. The rings typically contain small moons. It seems to be the gradual destruction of these by collisions that keeps the rings going for longer than would otherwise be predicted. Whether either the rings or the irregular satellites will persist over periods of a billion years depends on the ability of the Jovian planets to capture new material. The sizes of the rings and the numbers of irregular satellites will certainly fluctuate with time in the future. One particularly fascinating example is Neptune's main moon, Triton. It is an exception to the rule that large satellites move round their planet in the same direction as the planet spins. It is actually moving in the opposite direction, and as a result is gradually spiraling down toward Neptune. Before it gets there—in a few billion years' time—it will have a disastrous effect on all the moons lying between it and the planet. It will eventually fragment before it reaches Neptune, producing a magnificent, if relatively short-lived ring.

The regular satellites of the giant planets have histories that can be disentangled, like our own Moon's, by looking at the number of craters present. The majority have elderly surfaces in terms of the age of the solar system—meaning that they are now essentially inactive—but some still show signs of activity. The most extreme example is Jupiter's moon, Io, which orbits so close to Jupiter that the planet produces major tidal effects in the satellite. These heat the interior of Io, leading to numerous volcanoes on its surface. It has been estimated that the resultant volcanic flows are sufficient to cover Io with over a millimeter of new surface each year. A less obvious example of activity is provided by Saturn's largest moon, Titan. This, uniquely amongst satellites, has a thick atmosphere. But it has no magnetic field, and its orbit takes it outside the magnetosphere of Saturn. It is consequently exposed directly to the solar wind, and so should be losing material to space. The continuing existence of an atmosphere perhaps implies the continual addition of material from below. Further from the Sun still, Neptune's Triton seems to have a very young surface, so that continuing activity of some sort must still be at work there. (Observations of this distant moon are still very limited.)

The picture for the future is that all the regular satellites will undergo some change due to impacts on their surfaces, while a smaller number will continue to show significant activity on their surfaces. Some may even continue to be active over the Sun's lifetime on the main sequence. There will be drastic changes when the Sun becomes a red giant. Many of these satellites contain large amounts of ice. This is stable at present temperatures, but will melt and evaporate as the Sun gets hotter. For example, the three main satellites of Jupiter, circling outside Io's orbit, all contain considerable quantities of ice. Their surface temperatures when solar brightness reaches its peak may well exceed those currently experienced by Mercury. The consequent loss of all their volatile material will considerably reduce their size. Even as far out as Neptune's orbit, the temperature will reach a value similar to that at the Earth's orbit today, so all icy satellites will be affected.

There is one planet (now demoted from this status) that has not so far been mentioned—Pluto. Though it lies beyond the giant planets, it is very small. Indeed, its mass (even including the satellite that it possesses) is appreciably less than the mass of our own Moon. Its orbit too is peculiar, being so elongated that it sometimes comes within the orbit of Neptune. It is generally supposed that Pluto is simply a particularly large object from the Edgeworth–Kuiper belt that strayed inward. This interpretation has been bolstered by the recent discovery of similar-sized objects to Pluto that are orbiting further away in the belt. In addition, there is some reason for supposing that Triton is another large object from the belt that happened to pass close to Neptune and was captured. They, like other icy objects in the inner parts of the Kuiper belt, will be sufficiently warmed when the Sun becomes a red giant to achieve more Earth-like conditions. (To put it another way, the habitable zone round the Sun will by then have moved out toward the distance of the Kuiper belt.) But they are too small to develop significant atmospheres, and they will be particularly affected by the sweeping effect of the greatly enhanced solar wind.

Despite its oddity, Pluto's orbit round the Sun is reasonably stable, but the planet (along with its satellite) will almost certainly wander away when the Sun starts losing large amounts of its mass. In fact, this phase of the Sun's development presents a major problem to both the asteroid belt and the Kuiper belt. Material in both belts will spiral outward from the Sun along with all the planets. The changing orbits, both of the material in the belts and of the giant planets, will lead to greatly enhanced rates of collision as they try to adjust to the changing gravitational interactions. So the inner planets may be swallowed up, but the planets and other material further out can also expect to experience interesting times.

7. Our Galaxy

The Sun is one of about 100 billion stars that make up our Galaxy (written with a capital G to distinguish it from all the other galaxies in the universe). Seen from above, it looks a bit like a sombrero hat, with a brim consisting of a flattened disk of stars. These are revolving round a central crown, which is also made up of stars, though more densely packed than in the disk. Since the Galaxy looks the same from the other side too, its overall shape is more like two sombrero hats pressed together. A closer look reveals that the central bulge is not entirely symmetrical. It is actually slightly elongated—a little like the ball used in American football or British rugby, rather than a soccer ball. The bulge is some 20,000 light-years across, compared with the disk's 100,000 light-year diameter. (The exact size of the disk is difficult to measure because the stars tend to peter out toward its edge.) Orbiting the bulge are large groups of stars—spherical clusters—mostly formed from leftover bulge material in the early days of our Galaxy.

Note this word *light-year*. Now that we are talking about things outside the solar system, a light-year is the obvious unit to use. It is defined, unsurprisingly, as the distance that light travels in a year. But, equally, it is the distance that all radiation in the universe—radio waves, x-rays, and so on—travels in a year. It is the maximum speed at which communication can occur. For example, if the Sun suddenly ceased to exist, we on Earth would not know for just over 8 minutes, because the Earth is some 8 light-minutes from the Sun. We used a model in an earlier chapter which had the Sun the size of a grapefruit with an Earth the size of a flower seed circling it at a distance of 16 meters. On this scale, a light-year corresponds to about a thousand kilometers, and the nearest star to the Sun is some 4,300 kilometers away (roughly speaking, the distance from New York to San Francisco). In our part of the Galaxy—out in the disk—stars are typically a few light-years apart. In the bulge, they are closer together. There they orbit round a massive black hole which is situated at the center of the bulge.

Black holes have been talked about for a couple of centuries, but it is only in the last few years that they have actually been detected. Every object in the universe has a limiting speed associated with it. Unless you reach this speed, you can never escape from the object. (As we have seen, this "escape velocity" is just over 11 kilometers per second for the Earth.) How high the limit is depends on the gravitational pull of the object concerned. So to get away from the Sun requires a considerably higher speed than is needed to get away from the Earth. This idea of a limiting speed explains why black holes exist. Light can be thought of as a stream of particles moving at a particular speed—in this case, the extremely high speed of 300,000 kilometers per second. A body with a high enough gravitational pull could have this as its escape velocity, which would mean that light (and so all other sorts of radiation) could not escape from its surface. Such a body would be invisible—which is where the "black" comes from—but it would still have a gravitational pull that could be felt. Things could fall into it, but they could not come out again, since nothing can travel faster than the speed of light. That is where the "hole" comes from. The black hole at the center of our Galaxy is massive: it has something like 2.6 million times the mass of our Sun. So, although it cannot be seen, it has a considerable gravitational effect on the stars around it.

Though the disk extends a long way from the bulge, it is relatively thin—only 2,000 light-years or so in thickness. We are looking into this flattened disk when we sight along the Milky Way. (So our Galaxy is sometimes called the "Milky Way" galaxy.) Along with stars, the disk contains clouds of gas and dust. The stars currently contain much more of the material making up the material of our Galaxy than is contained in this *interstellar medium*. The clouds, and many of the stars (especially the brighter ones), are not distributed uniformly round the disk. Instead, they lie mainly along arms that spiral out from the region of the central bulge. As a result, the Galaxy, seen from above, looks rather like an octopus spinning round rapidly on a turntable. Its body corre-

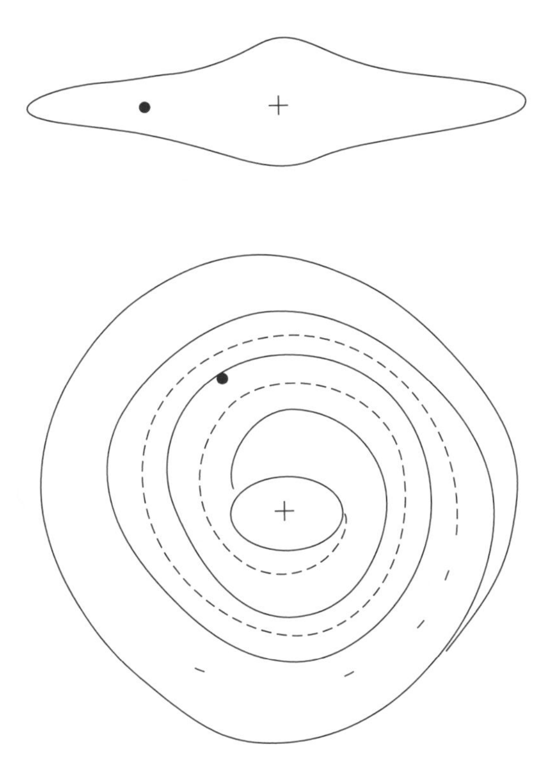

7.1 A diagram of our own Galaxy seen in cross-section and in plan view. The cross marks the center of the Galaxy, and the dot marks the position of the Sun.

sponds to the galactic bulge, and its tentacles trail out behind to the spiral arms. Because of this basic shape, our Galaxy is classified as a *spiral* galaxy. But since its central bulge seems to be elongated, it is often described more specifically as a *barred spiral.*

The Sun's Path

The Sun is a disk star, lying about midway between the galactic center and the edge of the disk. It follows a slightly elongated orbit round the center, so its precise distance changes with time. Compared with its current distance, it can reach a few hundred light-years closer to the center, or well over 3,000 light-years further away. The Sun is now near its closest approach to the center: it should reach there in about 15 million years time. In its present position, it is at the inner edge of a spiral arm (often unimaginatively called the *local arm*). The Sun takes about 240 million years to go round the galactic center once. This means that it has made the round trip some twenty times since the solar system was born. The estimated age of the Galaxy is well over 10 billion years, so over half of its history had passed before the solar system was born. Judging from what we can see of the rest of our Galaxy, the Sun seems to be a fairly typical star situated in a fairly typical part of the Galaxy. This means that analyzing the future of the Sun's local environment should give us a good idea of the future of the galactic disk as a whole.

The Sun, like other stars, bobs up and down through the galactic disk as it orbits round the center of the Galaxy. It can reach a maximum of 200–300 light-years above or below the central plane of the Milky Way as a result of this motion. The Sun last passed through the central plane quite recently—within the last 3 million years. It is currently on its way to its maximum height above the plane, which it will reach in about 15 million years from now. The galactic disk contains a lot of dust, which is a nuisance for observers since it acts as a kind of fog. The dust is strongly concentrated toward the galactic plane. As a consequence, from our present position near the plane, we have difficulty seeing through the fog to other parts of the Galaxy. This restricts our knowledge of what is happening there. In 15 million years we should have mounted above most of it, and will have the best possible view of other regions of our Galaxy. This will also be the time when the Sun will be closest to the galactic center, so our descendants will have a particularly good view of the galactic bulge.

The switchback motion of the Sun round the Galaxy does more than provide attractive views. Just as the Sun can raise tides in a planet, so the Galaxy can raise tides in the solar system. The size of the tide varies with the Sun's position. As the Sun oscillates up and down, the tidal pull of the Galaxy also varies. So far as the planets are concerned, this tide is too weak to be noticeable. For the comets in the Oort cloud, it is more important. They are far away from the center of the solar system, and the Sun's gravitational pull on them is small. Consequently, the galactic tide, though weak, can deflect them into new orbits. Some of these orbits may bring the comets in toward the Sun. In other words, the up-and-down motion of the Sun round the center of the Galaxy can lead to periodic increases and decreases in the number of long-period comets that we can see from the Earth. The Sun crosses the plane of the Milky Way, going upward or downward, every 30 million years or so. The number of comets coming into the inner solar system may therefore vary with this kind of period. There have been claims that such a variation with time can be found in the history of impact craters on the Earth's surface, but the evidence is, as yet, far from certain.

The Sun and Nearby Stars

A major problem in looking for evidence of regular changes with time in the number of comets hitting the Earth is that the Galaxy affects the Oort cloud in more ways than one. Our neighboring stars can also have an effect. The stars near the Sun all move round the galactic center in much the same way that the Sun does. But their orbits are not absolutely identical. If the Sun and nearby stars all moved round the center of the Galaxy in perfectly circular orbits, they would have a speed of around 200 kilometers per second. Because they mostly move in slightly elongated orbits, their speeds can differ from this by a few tens of kilometers. The region round the Sun is sometimes called the "local swimming pool." We know that any terrestrial swimming pool is being whirled round at high speed by the Earth's rotation. But, if we go swimming, what we actually see are simply the directions and speeds of all the swimmers in the pool relative to us. The same is true of the nearby stars: what we see is how they move relative to us.

Some of the stars around us move together in groups, while others (including the Sun) seem to be going their own individual ways. It is like a large airport where groups of tourists straggle hopefully around together while individual passengers thread their way between them. The constellation of the Great Bear provides a simple illustration of this. The central bright stars of this constellation are moving together as a group, but the stars at either end are separate individuals, which are going their own way. As a result, in 50,000 years time from now the constellation will look

7.2 The future movements of the brightest stars in the constellation of the Great Bear. The dots mark their present positions, while the ends of the arrows indicate where they will be in 50,000 years. The central stars are all moving together as a group, but the stars at either end are moving independently.

quite different. Scouts will no longer be able to use it to find the North Pole, because the two pointer stars will no longer point that way. What is true of the Great Bear is of course true of all the constellations we see. The patterns we trace now will change quite rapidly as the Sun and the nearby stars move relative to each other.

The fact that the stars in our vicinity are wandering about relative to each other raises a question. What is the probability that any two of them will collide? This depends mainly on how far apart the stars are. Our Galaxy includes some large clusters of stars. Near the centers of such clusters, the stars are quite close together. It is estimated that, for them, a collision might occur every 10,000 years or so. Correspondingly, the centers of these clusters contain a number of peculiar stars whose properties can best be explained in terms of collisions. For stars out in the spiral arms, the situation is markedly different. We are asking, in terms of our previous model, what is the likelihood that two grapefruit a few thousand kilometers apart will eventually hit each other? Clearly an individual star here represents a tiny target in the immensity of space. In consequence, the probability of a collision is remote: none is likely to have occurred in our vicinity since the Sun formed, nor is one likely to occur before the Sun reaches its dying phases. But this does not rule out "close" approaches between stars, so long as the word "close" is interpreted fairly liberally. If we set "close" to mean three light-years, then a star will wander to within this distance from the Sun every 100,000 years on average. The next major event of this sort should occur in 7,500 years from now, when a nearby star will pass at only 60,000 AU (about a light-year) from the Sun. This is far enough away to have little effect on the Sun and planets, but the star will actually pass through the outer parts of the Oort cloud. The resulting gravitational pull will send some comets out of the solar system altogether, but others will be diverted in toward the Sun. The sight may well be spectacular, with several comets visible from the Earth simultaneously (and an increased number of terrestrial impacts occurring too). The closest approach to the Sun by any other star over the next few billion years is likely to be at a distance of about 10,000 AU (though, in saying this, it has to be remembered that nearby faint stars are still being detected, so close approaches may be commoner than we think). At that kind of distance not only the Oort cloud, but the Kuiper belt too, will be disturbed. The spectacular cometary shower that will result may well increase the impact rate on the planets for a prolonged period after the actual encounter. From the viewpoint of a dweller on Earth the results will be spectacular, but unpleasant.

Interstellar Matter

Nor are the gravitational pulls of nearby stars and the tidal effect of the Galaxy the only two factors at work in the outer reaches of the solar system. Collisions between stars in our part of the Galaxy may be unlikely, but collisions between stars and interstellar clouds are virtually certain. The grandest of these clouds—called *giant molecular clouds*—may have masses equivalent to a million Suns and sizes up to 300 light-years across. The molecules referred to are mainly hydrogen gas, but the clouds also contain a considerable quantity of dust. The Sun is likely to encounter such a cloud once or twice every billion years. It will usually take a few million years to plow through it from one side to the other. During much of this time, the external world will disappear. The dust in the cloud acts as a particularly dense fog, blotting from sight anything in the universe outside the solar system. Giant molecular clouds come in various densities. It is the denser ones that are capable of producing the most significant effects within the solar system. The part that would be most affected is again the Oort cloud. The varying gravitational pull of the giant molecular cloud as the Sun moves through it disturbs the comets, especially the outermost ones. The result is again a major cometary shower into the inner reaches of the solar system, together with considerable loss of comets into interstellar space. So far as the total production of such cometary showers is concerned, galactic tides operate on an appreciably shorter timescale than giant molecular clouds, as do encounters with nearby stars. (Moreover, the number of giant molecular clouds in the Galaxy should diminish slowly with time as they are converted into stars.) Still, if we add together these three disturbing influences on the Oort cloud, their effects should be enough to ensure many showers of comets in the inner part of the solar system over the next few billion years.

The solar system is always moving through interstellar gas and dust. What varies is the amount of it in our vicinity. Electrically charged gas particles cannot penetrate into the inner parts of the solar system because of the solar wind blowing out from the Sun. Since the wind weakens with distance from the Sun, a standoff point is eventually reached. At this distance, the pressure due to the solar wind blowing outward is balanced by the pressure of the interstellar gas surrounding us. This boundary—the *heliopause*—is estimated to lie currently at some 100 AU from the Sun. The exact distance varies with the level of solar activity: when solar activity is low, the heliopause will creep in toward the Sun. Dust particles and gas that is not electrically charged can penetrate further into the solar system before they are stopped. At present, the amount of interstellar gas reaching Jupiter's orbit from outside the solar system is about equal to the amount of gas transported from the Sun to Jupiter by the solar wind. The Earth, being closer to the Sun than Jupiter, is better protected by the solar wind, so only a little of the interstellar gas penetrates inward as far as the Earth's orbit. This can change if the Sun's environment changes.

The solar system has just begun to enter a small interstellar cloud—sometimes called the "Local Fluff"—which is about 10 light-years across. It will take 200,000 years from now before we emerge from the other side of this cloud (because the Sun's speed relative to the cloud is low). The fringe of the cloud, where we are at present, has a fairly low density of gas and dust. As we near the center, the density of material will rise. It may become high enough for interstellar gas to penetrate as far as the Earth's orbit. The cloud seems to contain only a limited amount of dust, so even at its center we will still be able to see out to the rest of the Galaxy. When we next encounter a really dense cloud, not only will the view be blotted out, but also the Earth will be bathed for a long time in a much greater amount of interstellar material. Within such a cloud, the solar wind will be almost entirely suppressed, and the Earth will be surrounded by interstellar material. This may affect our upper atmosphere, and so what happens at the Earth's surface. It could, for example, change the ozone layer that currently protects us from ultraviolet light. It could also affect the amount of sunlight that the Earth receives. A significant change in this could lead to a major ice age. So the short-term future does not look too bad, but encounters with dense clouds may well affect our longer-term future.

The spiral arms of our Galaxy contain many small interstellar clouds such as the Local Fluff dotted at intervals. But, looking round our galactic neighborhood, it becomes obvious that there are also larger-scale structures. The interstellar medium near us has a bubble-like appearance. Each bubble is some 500 light-years across, with a skin of fairly dense interstellar material. The interiors are much emptier, though they contain small clouds of gas and dust, similar to the Local Fluff. The Sun itself lies in one of these bubbles—known, inevitably, as the Local Bubble. This is surrounded on three sides by larger bubbles labeled Loop I, Loop II and Loop III. (Though the naming sounds logical, Loop III actually lies in between Loops I and II.) The Local Bubble is about 300 light-years across, as measured in the plane of the Milky Way. This latter point needs to be specified because it represents the bubble's minimum size. Most interstellar gas and dust is concentrated into the plane, and this restricts the extent to which bubbles can grow sideways. Out of the plane, the pressures are much less, so the bubble can grow to a larger size in these directions. To an external observer, most bubbles would look like barrels sticking up through the Milky Way.

From the viewpoint of the solar system, Loop I is likely to have the most immediate impact on us. It is being driven outward from its center by the heat of the numerous bright stars that have

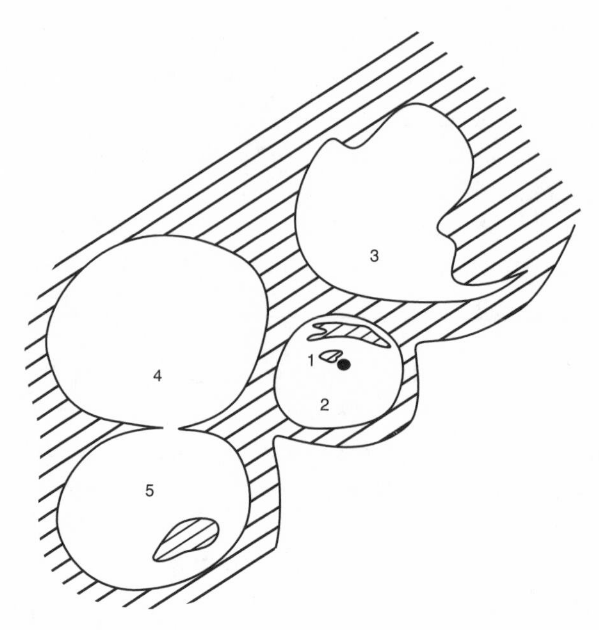

7.3 The galactic environment of the Sun. 1 Local Fluff; 2 Local Bubble; 3 Loop I; 4 Loop III; 5 Loop II. The Sun's position is marked with a dot. The shaded areas represent clouds of interstellar material.

formed within it over the last 10 million years or so. These stars are labeled the "Scorpius–Centaurus Association": the name is based on the constellations which frame them, as we look out from the Earth. The group lies a few hundred light-years away from us. Loop I, as it is blown outward by the Association, is pressing into the Local Bubble, sending interstellar material ahead of it into the Local Bubble's interior. The Local Fluff may prove to be the forerunner of still larger clouds yet to come. Having been in a region with little interstellar material for several million years, the Sun will, in the future, be spending a similar period of time in encounters with a series of interstellar clouds.

Supernovae

The driving force for the expansion of Loop I is connected with the many hot young stars it contains. Observation of other bubbles near us suggests that this cannot be the only explanation. Some bubbles seem to contain too few hot stars for this mechanism to work. There is another possible explanation. Stars a good deal more massive than the Sun evolve in a different way. Nuclear burning in their interiors continues to a much more advanced stage; in fact, successively heavier nuclei are created until nuclei of iron appear at the center of the star. At this point, further nuclear burning produces no extra energy. The star becomes unstable and explodes: it becomes a supernova. Its outer layers are blown off at high speed into space, sweeping up any material that lies in their way. The result is an expanding sphere of gas, similar to the bubbles we see around us. Sometimes the explosion leaves behind a highly compressed neutron star. As we saw earlier, a white dwarf contains a mass similar to the Sun's compacted into a star the size of the Earth. A neutron star also has a mass similar to the Sun's, but now contained within a body only about 20 kilometers across. If a large remnant is left from the explosion, it can even compact itself sufficiently to become a black hole. This will be minute compared with the black hole at the center of our Galaxy, but still capable of attracting any nearby material.

Massive stars not only evolve through more stages than the Sun, they also evolve much more rapidly. They reach the supernova stage in times measured in millions, or tens of millions of years. A group of bright young stars like the Scorpius–Centaurus Association can be expected to produce at least one supernova during its early lifetime. In this case, the energy from the supernova and the heat from the stars reinforce each other to create an expanding bubble of interstellar material. In other cases, a supernova may be sufficient by itself to create a bubble. It is generally believed, for example, that the Local Bubble was created by a supernova explosion. The material in the skin of our bubble is currently being slowed down as it encounters the skins of other more active bubbles.

Supernovae actually come in two varieties. The young ones, just described, are called Type II supernovae. Type I supernovae result from the explosion of much older stars. Many of the stars in our Galaxy are actually double—two stars revolving round each other. If the two stars have different masses, they will evolve at different rates. In particular, one star may reach the white-dwarf stage while the other is still in the middle of its evolution. For stars close together, this can lead to a problem. The star that is still evolving may, when it expands, encroach on the space occupied by the white dwarf. The problem, as we have seen, is that degenerate matter, such as that in white dwarfs, has a limit to how much material can be loaded on to it. If the companion star feeds it too much, the white dwarf will ultimately explode. This is what produces a Type I supernova. Because new stars are to be found mainly in the spiral arms, the bubbles around us are associated with Type II supernovae. Type I supernovae occur between the arms, as well as within them.

The material shot out by supernovae of both types moves away from the explosion at high speeds—perhaps 10,000 kilometers per second. It is then gradually slowed down by collisions with interstellar clouds. But a small part of the material is shot out at even higher speeds, and these particles may roam far away from their original source. Such particles are called *cosmic rays*. (Supernovae are not the only source of cosmic rays, but they are the main cause of sudden injections of such particles into our Galaxy.) The high speeds of cosmic rays can carry them for considerable distances throughout the Galaxy. Since they are electrically charged, they can be diverted by strong magnetic fields, but otherwise they keep going until something gets in the way.

On Earth, the terrestrial atmosphere and the magnetic fields of both the Earth and the Sun act to protect us from the direct effects of cosmic rays. How well they can protect us depends on how close the supernova explosion is to us. The particles and radiation from a nearby supernova can produce significant, though temporary, changes in the Earth's atmosphere. For example, they can interact with the Earth's upper atmosphere, altering the ozone layer and allowing through a flood of ultraviolet light. It also seems that an abundance of cosmic rays can affect the greenhouse effect—in the direction of lowering temperatures at the Earth's surface. If a nearby supernova occurs while the Sun is passing through a large interstellar cloud—quite possible, since massive stars are formed in such clouds—the effects will be appreciably worse. As we have seen, such a cloud suppresses the protection we currently enjoy from the Sun, allowing cosmic rays from the supernova to make an even greater impact on the terrestrial environment.

The interstellar material in our local part of the Galaxy has clearly been strongly affected by a number of supernova explosions. Something like twenty supernovae have occurred within 350 light-years of the Sun during the past 10–20 million years. Roughly speaking, any supernova that occurs within 150–200 light-years of the Sun is likely to have a noticeable effect on the solar system. The Scorpius-Centaurus Association in Loop I is now over 400 light-years away, but it is moving relative to the Sun. Looking back into the past, it came to within 130 light-years of us about 2 million years ago. A puzzling feature of recent deposits in the Earth's crust is that they contain radioactive iron, which seems to date back to about 2 million years ago. It has been suggested that this radioactivity was produced by the explosion of a nearby supernova, probably in the Scorpius–Centaurus Association when it was near to the Sun. Such a close explosion would produce other effects, and indeed 2 million years ago also saw a significant extinction of living forms on Earth. So, to our list of future problems for life on Earth, we must certainly add nearby supernovae. We can get some idea of the timescales involved by studying how cosmic rays have affected meteorites out in space. Examination of the radioactivity it has produced in them suggests that cosmic-ray bombardment of the solar system becomes stronger every 100–200 million years. To see whether this will continue in the future, we must consider why these variations occur.

Spiral Arms

The Sun is currently positioned near the edge of a spiral arm, on the side toward the galactic center. Spiral "arms" are basically waves—like ripples on water—that sweep round the disk of the Galaxy at fairly regular intervals. When the wave encounters large clouds of gas, it compresses them, so encouraging the formation of new stars. As a result, large groups of hot young stars—such as the Scorpius–Centaurus Association—can be found clustered along the spiral arms. The local arm and the Sun are moving at different speeds, so the Sun is not a permanent member of the arm. It follows its own orbit round the galactic center, while, at intervals, the spiral wave sweeps past it. So the Sun is sometimes embedded in a spiral arm and sometimes not.

Encounters between the Sun and a spiral arm occur every few hundred million years. Because of an interesting coincidence, the exact period is difficult to determine. The length of time the Sun takes to orbit the galactic center turns out to be similar to the length of the time the spiral-arm waves take to make one circuit. If it becomes embedded in a spiral arm, the Sun stays there longer than stars that are closer to the galactic center, or further away, because they are moving at different speeds. But the fact that other stars spend less time in the spiral arms than the Sun, also means that they encounter the arms more frequently. For the low relative speed of the Sun and the spiral arms means that the Sun not only remains in the arm regions for longer, but also that it remains in the inter-arm regions for longer. Since the inter-arm regions take up more space in our Galaxy than the arms, the Sun spends appreciably longer periods outside spiral arms than within them. This brings us back at last to the question of supernovae. The Sun's orbit provides some protection for the solar system against receiving too many cosmic rays. We will be out of the arms, where Type II supernovae occur, more than most stars. Even so, life between the arms is not entirely free of hazard. Type I supernovae can be found as readily in the space between arms as within an arm. Indeed, there is a suspicious-looking white dwarf plus companion star only 150 light-years away from the Sun. If this system exploded now as a Type I supernova, the results for us would be devastating. Fortunately, it still has some time to go before it reaches that critical state. Since the Sun is moving relative to the system, we should be a good long distance away when it finally blows up.

How near the Sun ventures to newly forming stars will vary from one passage through a spiral arm to the next. Big interstellar clouds, where new stars preferentially form, are often fairly isolated from each other. The spaces between them are usually several times the size of the individual clouds. Consequently the Sun may only encounter a big cloud once in every 5–10 spiral-arm passages, so it is difficult to guess what any individual passage will bring in terms of cosmic rays. Still, looking to the future, we can predict that the Earth will face increased exposure to cosmic rays every few hundred million years or so–in reasonable agreement with the past frequency measured from meteorites. But we can now add that, once or twice in the next billion years, the solar system can expect to receive a really massive dose of cosmic rays.

The overall pattern of spiral arms is determined by large-scale waves in the Galaxy, but the characteristics of an individual spiral arm depend on what is happening in its local environment. Within a spiral arm, the main features of the interstellar gas are—as we have seen—typically dominated by the heat from hot young stars, and especially by the effects of supernovae. Older stars like the Sun dodge in and out of these various features. But this picture obviously raises a question regarding the future. If large clouds are transforming their contents into stars every time a spiral wave hits them, then the amount of interstellar material in the Galaxy must be continually decreasing. When will the Galaxy run out of star-forming material?

Most of the known giant molecular clouds lie either near the distance of the Sun's orbit from the galactic center, or closer in toward the galactic bulge. Few have been spotted in the outer parts of the disk. So the formation of hot new stars is currently confined mainly to the more central regions of the disk. In our local arm, interstellar matter makes up somewhere between 15–30 percent of the total mass present. This figure indicates that much of the interstellar material in our part of the Galaxy has already been converted into stars. So far as we can judge, this is not too bad an estimate for the inner arms of the Galaxy as a whole. It follows that, if we can decide how rapidly stars are being formed now, we can work out roughly how much time will pass before all the interstellar clouds will be turned to stars.

Unfortunately there is a complication. Stars, especially toward the end of their careers, recycle some of their contents back into the interstellar medium. For massive stars, this occurs mainly via supernovae explosions. For stars in the same mass range as the Sun, it is via the more moderate processes that act from the red-giant stage onward. This replenishment of interstellar material happens on different timescales. For a massive star, the time can be measured in tens of millions of years. For a star such as the Sun, it is measured in billions of years. Since both types of star are being created all the time, working out just how much material will be recycled at any particular point in the future is far from easy. In terms of current conditions, the amount of material being returned to the interstellar medium seems to be something like a fifth of the amount that is condensing into stars. If we accept this estimate, most of the interstellar material will have disappeared by a few billion years from now. At that stage, lower-mass stars will still be returning some material to space, so star formation will continue, but the heyday of the massive stars will be over. This picture fits in well with observations that the peak period for star formation in the Galaxy has passed. Estimates vary, but it may be that our Sun formed toward the end of that peak period. However, this assumes that, when you look at the Galaxy, what you see is what you get. In other words, the Galaxy is a closed box—nothing is added from the outside and nothing is lost. As we shall see in the next chapter, this is untrue. Our Galaxy interacts and exchanges material with the outside world. This can have a major effect on how exactly our Galaxy develops, but cannot alter the fact that ultimately all the star-forming material will be used up.

The Evolving Galaxy

The material that is ejected from stars is not the same chemically as the material from which they originally formed. Some of it has been involved in the nuclear-burning processes which have kept the star alight. Consequently, the material thrown out contains more of the elements heavier than hydrogen and helium than the original interstellar material did. To put it another way, the recycling of material by stars continually increases the proportion of heavier elements in the interstellar medium. So stars being born now contain more of the heavier elements than the Sun does (and the Sun, correspondingly, contains more of those elements than stars born earlier in the lifetime of the Galaxy). The extent of the reprocessing depends on the rate of star formation, especially of the massive stars that burn up and return material to space quickly. For our Galaxy as a whole, this means that the accumulation of heavier elements is going on most rapidly in the spiral arms at a distance of less than the Sun's orbit from the center. This leads to a gradient of heavier element pollution, going from a maximum in toward the central bulge to low values in the outer parts of the Galaxy.

Near the center itself, observations suggest that a burst of rapid star formation will occur in about 300 million years from now. The number of hot young stars created will be enough to lead to a supernova explosion there every year. As it happens, the central region seems to have experienced a supernova explosion quite recently—within the last 50,000 years—so that the galactic center is currently surrounded by a bubble of expanding gas. Strong emissions of material are also coming from the central black hole. At first sight, this might seem odd. If a black hole can absorb material but not eject it, how can it make material around it expand? The answer lies in its voracious appetite. As the attracted material spirals inward toward a black hole, it jostles violently together. (Tidal forces near a black hole are high. So anything of any size, such as a star, gets shredded to pieces as it approaches the hole.) The picture is rather like water going down the plughole in a bath. The water piles up, spinning rapidly as it tries to force its way down the outlet. The pileup round the black hole heats the material, and leads to explosive outbursts that can power an expansion outward. One part of the circling material falls into the black hole, while another part is shot outward. The speeds involved are high. Near a black hole material can be moving at an appreciable fraction of the speed of light. The central black hole in our Galaxy is currently consuming material at a relatively modest rate. This may be because the nearby supernova mentioned above has swept away some of its potential food. Even so, there is some evidence that, as little as 350 years ago, the region round the black hole was a million times brighter than it is at present. For the future, we can clearly expect spasmodic outbursts round the center as the black hole there is alternately gorged and starved by the amount of surrounding material. These outbursts will occur on timescales ranging from hundreds to hundreds of millions of years, with the longer-term ones likely to be the larger in scale. All this activity near the center may have appreciable effects within the inner regions of the galactic disk, but it is unlikely to worry us at the Sun's distance.

Only the more massive stars shoot much material back into space. Stars with a mass about a third that of the Sun or less never reach a high enough central temperature to ignite helium. They never become red giants, and so never lose material at that stage. After their main-sequence career, such stars—known as *red dwarfs*—move on directly to the white-dwarf stage. As we have seen, the central regions of the Sun do not mix with its outer layers, so nuclear burning only affects the hydrogen round the center. In stars of lower mass, the convection in the outer parts reaches right down into the central regions. As hydrogen is burnt, the helium residue is carried up to the surface, while new hydrogen is brought down to the center. This means that the whole hydrogen reservoir in the star is available for burning. In addition, these smaller, fainter stars burn hydrogen much more slowly than the Sun does. The overall result is that red dwarfs can have enormously long lifetimes—10 trillion years or more (where a trillion means a billion billion).

There is a lower limit for red dwarfs as well as an upper limit. Below about 8 percent of the Sun's mass the central temperature cannot rise high enough even to ignite hydrogen. The process of contraction produces heat, but that is the only source of energy. Objects of this sort are labeled *brown dwarfs*. Below about 1 percent of the Sun's mass, the heat generated in this way is so small that the object can reasonably be called a planet. But that is not quite all the story. The figures quoted are for stars which have the current galactic chemical composition. As we have just seen, future stars will be born with a greater proportion of heavier elements. It turns out that this will make it easier for low-mass stars to burn hydrogen, and so take their place on the main sequence. It follows that red-dwarf stars will form even more readily in the future. Bright massive stars may dominate activity in the Galaxy now, but they are few in number. The great majority of stars in the Galaxy are either red dwarfs or brown dwarfs. Looking to the future, as new material for forming stars becomes scarcer and scarcer, it seems that these stars must dominate. However, we must hold over a final decision on when this will occur until we have looked at our Galaxy's external environment in the next chapter.

And What About Life?

Comparing the Earth with the other planets gives some idea of what has been required for the development of life in our solar system. Can a comparison of our Sun with other stars correspondingly give clues to the likelihood of life in the Galaxy as a whole? Liquid water seems to be a prime requirement for life. This means that the central star must have a reasonably extended zone round it where a circling planet can remain at the necessary temperature for liquid water. For stars considerably cooler than the Sun, this habitable zone is narrow. Stars much hotter than the Sun face a different problem. Judging from the Earth's history, it may have taken a billion years for simple life forms to develop. Other stars—even presuming they possess suitable planets—must therefore not change too much over this sort of time if they are to support life. Stars appreciably hotter than the Sun pass through their lifetime on the main sequence more quickly. They change appreciably in less than a billion years. It follows that stars like the Sun—not much hotter, nor much cooler—provide the best environment for life.

Our existence on Earth has depended not only on its supply of liquid water, but also on the fact that it is a small rocky planet. (It has a convenient surface for evolutionary experiments, with a gravitational pull that is neither too big nor too small.) The Earth exists because the interstellar cloud from which the Sun and the Earth formed had an adequate supply of heavier elements to provide the rocks. Though the Sun consists mainly of hydrogen and helium, its supply of heavier elements is, by stellar standards, quite reasonable. Stars born from gas clouds containing fewer of the heavier elements would find it difficult to accumulate rocky planets round them. Equally, a star with a much higher proportion of heavy elements than the Sun might form large rocky planets rather than small ones. The Sun, once again, has got it just about right to provide us with an acceptable home.

Since star formation has gone on more vigorously in the inner regions of the Galaxy, this suggests that planetary systems are more likely to be found in that direction. But this poses a problem. The more stars there are in the vicinity, the more a planetary system encounters disturbances. Nearby supernovae, stars passing close by, and giant molecular clouds will all be encountered more frequently in the inner parts of the galactic disk. None of these make conditions easier for the creation of life. The virtue of the Sun's orbit round the galactic center is that it keeps such disturbances down to an acceptable level. The galactic habitable zone where this advantage can be enjoyed is estimated to cover the region between about 20,000–30,000 light-years from the galactic center. The stars which are likely to have small rocky planets attached to them in this zone mainly have ages in the range of 4–8 billion years. There may, of course, also be suitable planetary systems round stars either closer to the galactic center, or further away from it, than the Sun. But these other regions of the Galaxy provide a less helpful environment for the evolution of life.

Even with all these restrictions, there should be plenty of Earth-like planets circling their suns in acceptable environments. Indeed, current studies are showing that many of the stars in the Sun's neighborhood have accompanying planets (though the observations are not quite good enough yet to look for Earth-like bodies in appropriate orbits). If, as is generally assumed by biologists, life will develop whenever conditions are appropriate, then some planets around other stars should surely evolve life. Impacts must occur in other planetary systems as they do in ours. So small amounts of material will be redistributed from planet to planet over a long enough period of time. Correspondingly, if primitive life forms appear on one planet, they should appear on any other planet with suitable conditions. But this only applies within a single planetary system. Even primitive forms of life are unlikely to survive a journey between the stars, and in any case the likelihood of them encountering a suitable environment at the other end is exceedingly small. So each planetary system will have to develop its own life forms. Detecting the existence of such life will not be easy. The most obvious test is to look for planetary atmospheres with an unusual chemical composition. Thus, to an external observer, the Earth would stand out in the solar system because it is the only planet with an atmosphere of nitrogen and oxygen.

Detection of life elsewhere should be easier if the primitive life forms eventually evolve to produce intelligent life. One noticeable characteristic of the human race is its urge to communicate. For decades past, as noted in the introduction, some radio signals—not least television broadcasts—have been leaking out into space. Any intelligent life within 75 light-years of the Sun can now receive signals from us. But here we encounter a difficulty first pointed out many years ago by the physicist Enrico Fermi: "If there are so many people out there, where are they?" The problem is that we have not detected similar information-laden signals from the vicinity of other stars. Since some stars are older than the Earth, there has been ample time for intelligent life to appear there before it developed on Earth. Correspondingly, there has been plenty of time available for signals from such life to pervade the Galaxy. Yet despite continuing efforts to detect extra-terrestrial signals—particularly through the SETI (Search for Extra-Terrestrial Life) program—nothing like an alien message has been discovered. Possible reasons will be examined in the final chapter, but this silence certainly poses a challenge to how we view the development of intelligent life, and perhaps of life itself, elsewhere in the Galaxy.

To recap—the Sun seems to be the right sort of star, born at the right sort of place and time, and with the right sort of galactic orbit, to nurture not only life but even intelligent life. With this as our starting point, we can ask: what are the prospects for planets developing life in the future? Our Galaxy is gradually running out of interstellar gas. The overall rate of star formation is therefore falling: the peak period of star formation occurred long in the past. This means that disturbing events, such as supernovae explosions, are also becoming less frequent. So the inner parts of the Galaxy are becoming more environmentally friendly for life as time passes. But it seems that the increased amount of heavier elements in the inner parts of the Galaxy may become too large to allow small rocky planets to form in nice circular orbits round their central star. Gas clouds further away from the galactic center than the Sun will gradually increase their heavy element content in the future. This should bring them up to the point where, as with the Sun, small rocky planets can readily form. The problem will continue to be that, because the amount of interstellar gas in the outer reaches is limited, the number of suitable stars forming there will also be restricted. Perhaps for the really long-term future of life we need to look more widely. We will return to this question in the final chapter.

8. Galaxies

Galaxies, like stars, come in a range of sizes. Like stars too, there are not so many big galaxies, but plenty of small ones; and again, like stars, the big ones are called "giants" and the small ones "dwarfs." But photographs of galaxies are much more interesting than photographs of stars because galaxies have such a variety of shapes. The main division is between spiral galaxies like our own and elliptical galaxies. Spirals come in two basic forms, depending on the nature of their central bulge. Ordinary spirals have a fairly symmetrical central bulge; barred spirals have an elongated central bulge. The relative size of the bulge and the spiral arms varies, as does the tightness with which the arms are wrapped round the bulge. Given all these possibilities, it is hardly surprising that spiral galaxies can differ considerably from each other in appearance. Elliptical galaxies differ less. They look more like the central bulge of a spiral galaxy, but less flattened and without the spiral arms. As their name suggests, they are more or less elongated in cross-section. Besides these main types, some galaxies are labeled "irregular"—meaning that they do not have an obvious shape—while others are "peculiar"—meaning that they may have a definite shape, but there is something odd about them.

The important thing when we examined the future of stars was to determine how they evolved with time. That made it possible to predict what would happen to the Sun and other stars in the future. We must obviously ask a similar question about the evolution of galaxies. In particular, are all the various sorts we see simply different stages in the evolution of individual galaxies, or does each type of galaxy have its own separate evolutionary path? Unfortunately, there is a major complication here. In the last chapter, we discussed the future of our own Galaxy as though it

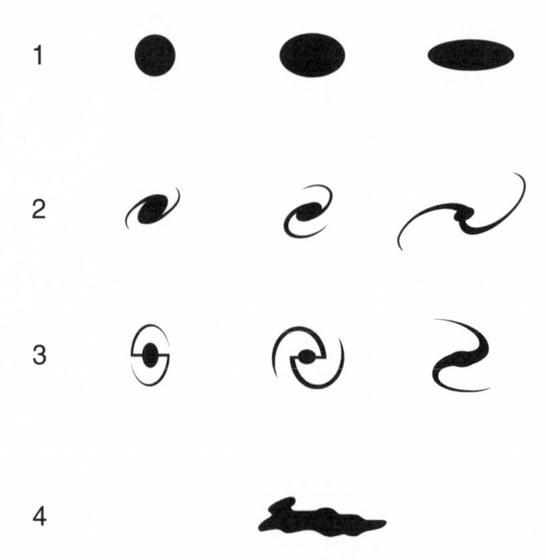

8.1 The main types of galaxy and the various forms in which they come. 1 elliptical; 2 spiral; 3 barred spiral; 4 irregular.

was completely separate from anything else in the universe. But galaxies are typically surrounded by other galaxies. Can the influence of these really be ignored?

The Galactic Halo

The description of our Galaxy in the last chapter left out one part of its structure. Around both the central bulge and the disk, there is an extended, though low-density, halo of gas (with some old stars embedded in it). This halo plays a significant role in the present discussion of galactic environment, for it has been found to contain denser clouds of material which are falling inward to the central plane of the Galaxy. The clouds consist mainly of hydrogen, and may be as massive as 10 million Suns. Taken together, they are equivalent to about a tenth of the total interstellar gas in the spiral arms. Some of the clouds have originated near the galactic plane. We saw in the last chapter that the material ejected from supernovae can expand more easily upward and downward out of the galactic disk. As a result, some of the ejected gas goes all the way out into the halo, where it forms clouds. What goes up, eventually comes down: these clouds ultimately fall back and rejoin the main part of the Galaxy. This recycled material can then go on to form new stars. In terms of evolution, such recycling has a limited impact. Excursions of material out into the halo may slow down the rate of formation of new stars, but do not alter the overall picture of our Galaxy's future as a place where star formation decreases with time.

But not all the clouds in the halo have originated in the disk. Some of the infalling clouds appear to be coming from outside the Galaxy. There are four possible sources. The first is simply gas that has been ejected from our Galaxy at such a speed that it has actually wandered off into intergalactic space before coming home again. Then there is material left over from the formation of the Galaxy. Like stars, galaxies form from collapsing clouds of gas. In both cases, material out on the fringes may lag behind, so that the central body forms without it. In the solar system, for example, we have the comets left behind in the Oort cloud. This leftover material may now be belatedly falling inward to add itself to the Galaxy. Thirdly, just as there is diffuse material—the interstellar medium—lying between the stars in our Galaxy, so there is a diffuse intergalactic medium between our Galaxy and other nearby galaxies, which may be adding to the infall of material into our Galaxy. The final source—material captured from nearby galaxies—will be looked at in the next section. In terms of its effects on future development, the first type of gas cloud is similar to material recycled through the corona. Its long absence simply slows the rate of star formation. The other two types represent the addition of new material to the Galaxy. Their capture increases the number of stars that can be created in the Galaxy, as well as extending the future period of star formation. In addition, recycled material will be more contaminated with heavier elements than the new material, so there will be chemical differences between stars formed from the two sources. But the influence of these sources does not introduce a fundamental change in the way that our Galaxy will develop in the future.

The Local Group

If galaxies had a theme tune, it would probably be "You'll never walk alone." When you look out beyond our Galaxy, you can certainly see isolated galaxies. But they are not the dominant feature. What catches the eye is the grouping of galaxies into clusters. Such clusters come in all sizes from the very big, containing many thousands of galaxies, to the very small, with just a handful of galaxies. As with most things in the universe, the small considerably outnumber the big. So it is no surprise to find that our own Galaxy is a member of a small cluster. Nor is it surprising that astronomers use their favorite word "local" to describe it: we are, they say, a member of the *Local Group* of galaxies. "Local," in this case, means that any galaxy within about 4 million light-years of us is probably a member of our cluster.

The Local Group contains 30–40 member galaxies. The exact number is difficult to determine, partly because most are small and faint, and partly because some may lie hidden behind the obscuring dust clouds in our own Galaxy. For once, we do well in terms of the grandness of our home. Our own Milky Way galaxy is one of the two brightest and biggest objects in the Local Group. The other is the Andromeda galaxy (called after the constellation in which we see it). It too is a spiral galaxy, though apparently not barred like ours. There is one other fairly bright spiral in our cluster; the remaining galaxies include ellipticals and irregulars (plus dwarf galaxies whose shapes are difficult to determine). After the three spiral galaxies, the next brightest galaxies in the Local Group are two irregulars, called the Large and Small Magellanic Clouds. They are visible from the southern hemisphere of the Earth, where they lie quite close to the band of the Milky Way. (Their name is a memento of Magellan's voyages southward around the year 1500.) In this case, their appearance of being close to the Milky Way is actually true: the two galaxies are near neighbors of ours in the Local Group. The Large Magellanic Cloud is less than 200,000 light-years away, and contains about a tenth as many stars as the Milky Way. The Small Magellanic Cloud is a little more distant, and contains less than a quarter as many stars as its companion. (Compare these distances with that of our sister spiral in Andromeda, which is well over 2 million light-years away.)

These distances are worth thinking about. When we discussed the evolution of stars, we mostly ignored the possibility of stars colliding. The reason was that stars are small objects and, in the spiral arms at least, their distances apart are hugely greater than their sizes. Consequently, the likelihood of collisions between them is remote. This argument does not hold for galaxies. Galaxies are large, and their distances apart in clusters are not vastly greater than their sizes. (For example, the distance to the Magellanic Clouds is only about twice the diameter of our own Galaxy.) This means that collisions or near-misses are quite likely to occur, and can have a major effect on the way in which a galaxy evolves. To see what kind of things might happen, consider the progress of the Magellanic Clouds.

It is calculated that both Clouds were close to the Andromeda galaxy some 10 billion years ago. They escaped that encounter, only to be captured by the gravitational pull of our own Galaxy, perhaps 6 billion years ago. As they approached the Milky Way, gravitational interaction with our Galaxy—beginning, say, 1.5 billion years ago—led to some material escaping from the Clouds. This now forms a long plume—called the "Magellanic Stream"—lying along the paths of the two galaxies. Further breakup occurred when the Large and Small Clouds passed close by each other some 200 million years ago. (The Small Magellanic Cloud has obviously found it hard to survive all this interaction: it gives the impression that it is falling to pieces.)

The Magellanic Clouds are now satellites of our own Galaxy. Like any satellite, they have an effect on the host that they orbit. Our satellite, the Moon, raises tides in the Earth. Similarly the Clouds raise tides in the disk of our Galaxy. Looking forward in time, the gravitational interaction between the Milky Way and the Magellanic Clouds will almost certainly lead to the latter colliding with our Galaxy some time during the next several billion years. What will happen then? The distances between stars in the Magellanic Clouds are similar to those in the spiral arms of our Galaxy. Consequently, collisions between our stars and their stars are very unlikely (unless the Clouds hit the central bulge, rather than the arms). In principle, the stars in the Magellanic Clouds could pass straight through the spiral arms of the Milky Way, and the main thing we would notice would be a doubling in the number of stars around us. In practice, the gravitational interactions make things more complicated. The relative speeds of the Magellanic Clouds and our own Galaxy are sufficiently low that the material in the Clouds should be captured and merge with the Milky Way, rather than pass through it. The main gravitational pull of our Galaxy comes from the bulge in the center, so it strongly influences the motion of bodies captured from outside. The Magellanic Clouds will probably collide with the disk of our Galaxy, rather than the bulge. But the gravitational pull of the bulge will then ensure that most of the captured stars start moving round the center of our Galaxy. In the process, the orbits of stars currently in the Galaxy could be heavily disturbed. Consequently, by the time that the Sun reaches its final phases, it may well be following a considerably different path round the Galaxy from now.

This comparatively peaceful picture does not apply to the interstellar clouds present. Because the clouds are so large, they cannot avoid hitting each other when the galaxies collide. The impacts will compress the clouds and heat them up. This in turn will lead to the rapid formation of a whole new generation of stars. As compared with the present rate of star formation in our Galaxy, things will speed up by a factor of 10–100 during this starburst period. Some of the newly created stars will be massive and will quickly explode as supernovae. Part of the ejected matter will stay in the disk, but part will be recycled through the halo, so there

8.2 The distribution of the brighter galaxies in the Local Group. 1 our own Galaxy; 2 the Andromeda galaxy; 3 the Large and Small Magellanic Clouds.

will be a continuing period of star formation after the initial burst. Wherever the Sun is in the Milky Way, it will be treated to a remarkable fireworks display. If it is unlucky it will be nearby, and the solar system will be bathed in the material ejected from the supernovae—another good reason for not being around then.

Galactic Harassment and Cannibalism

As the lengthy Magellanic Stream shows, the Clouds are already losing gas to the halo of our Galaxy. In fact, any smaller galaxy unfortunate to enough to interact with our own can lose considerable material, even if it is not gulped down whole. The milder forms of mistreatment are often referred to as "galactic harassment," whereas total removal is "galactic cannibalism." The result in either case is to add material to our Galaxy. For galaxies like the Magellanic Clouds, which contain considerable amounts of interstellar gas, this can be the most important way of introducing new halo material. The amount added varies with time, depending on the orbit of the satellite galaxy round our own. For one in a close orbit, loss can occur for much of the time. For instance, the Sagittarius dwarf galaxy (so-called because we see it in the constellation of Sagittarius) lies some 50,000 light-years away from the Sun on the other side of the central bulge. It is a satellite of our own Galaxy and is losing matter as it orbits through the halo, leaving a stream of stars behind it. Eventually, this dwarf galaxy will be totally pulled to pieces, and the stream it leaves will be absorbed into the Galaxy, though it may take some billions of years before this is finally achieved. Looking at the region of the Galaxy near the Sun, it is possible to see streams of stars which move with different speeds and directions from the majority of the stars in our neighborhood. For example, the nearby red-giant star Arcturus is the brightest member of one such stream. It has been suggested that such streams represent the remnants of stars captured from other galaxies, which have not yet become fully reconciled with their new home. Such alien stars may have different chemical compositions from the home-grown stars in their vicinity. Their presence makes it more difficult to determine the history of our Galaxy—as we tried to do in the last chapter—by studying how the chemical composition of its stars varies with distance from the central bulge.

Clearly, in the competition for acquiring new material, the Galaxy is winning out at the expense of its smaller neighbors. In the process, it is not just consuming smaller galaxies, but is also changing their nature. For example, the Local Group contains a number of dwarf irregular galaxies. These contain significant amounts of interstellar gas from which stars are still being formed. The interesting thing is that they are mainly situated in the outer parts of the Group. Near to the Milky Way, dwarf galaxies seem to be predominantly elliptical or spherical in shape. The link seems to be that dwarf irregulars, if they venture too close to our Galaxy, have a significant amount of their interstellar medium stripped away. As a result, star formation ceases, and they are transformed into dwarf ellipticals. As time passes, more and more small members of the Local Group will be subject to this stripping process.

The Galaxy is one of the two heavyweight members of the Local Group. The other—the Andromeda galaxy—is often referred to as its twin, though the two are not entirely identical. But all that has been said so far about the Milky Way is also true of the Andromeda spiral. In particular, it is surrounded by an extended halo, and is stripping material from nearby dwarf galaxies. The future development of the Local Group seems obvious. The two main galaxies are gradually pulling in the remaining material in the Local Group and absorbing it into themselves. This leaves the major question of what will happen to these two galaxies. Measurements of the Andromeda galaxy show that it is approaching us at a speed of a few hundred kilometers per second. It seems set on a collision course with our Galaxy. What happens in the future will depend on the details of the actual collision—central or glancing, flat-on or edge-on? As with the Magellanic Clouds, there will be a spectacular flare-up when the interstellar clouds in each galaxy collide. In addition, the haloes of the two galaxies will collide and their central bulges will mingle. The stars present will pass each other by, and the two galaxies will swing backward and forward through each other, for hundreds of millions of years before everything settles down.

The combined galaxy will have two black holes at its center. They too are likely to swing round each other, perhaps for hundreds of millions of years, before they merge in a huge outburst of radiation.

Calculations suggest that the final outcome will be the formation of a giant elliptical galaxy with a single massive black hole at its center. There will be no more Milky Way. The disk in which the Sun currently resides will have disappeared, as will the interstellar clouds that currently hinder our ability to see the rest of the Galaxy. In its place, the view from the solar system will be of stars stretching out rank on rank in all directions. This supposes, of course, that the Sun will still be around. Though collisions between stars in the spiral arms can be ignored, the size of the gravitational interaction between the two galaxies is such that some stars may be thrown out of the combining galaxies altogether. There is an outside chance that the Sun may be one of these. If it is, the Sun will wander off on a grand tour of its own (though it may well eventually return home again). Alternatively, the Sun might be thrown inward, in which case there is a slight chance that it will end up as food for the central black hole.

So when is this transformation likely to occur? The major problem is uncertainty about how exactly the Andromeda galaxy is moving relative to our own Galaxy. It is certainly coming toward us, but it is not clear whether it is approaching head-on or at an angle. If the latter, then the Andromeda galaxy may swoop round us, only to disappear again into space: rather as a comet swoops in toward the Sun and then recedes. There will certainly be tidal effects as the two galaxies interact, with stars and gas being pulled out from both, but they may not fuse into a single galaxy. The close encounter could be as soon as 3 billion years from now, though it will probably take a bit longer. At that stage, the Sun will still be a main-sequence star and the solar system will be intact. If the result is a near-miss rather than a collision, the two galaxies will continue circling each other; but sooner or later—on a timescale of several billion years—a merger will inevitably happen. But the picture of the Local Group as a single large elliptical galaxy may not be the final end-product. If the merged galaxies were subsequently to encounter a large amount of new material (from intergalactic space or from other galaxies), it might be possible for a new disk to form. In such a case, the Sun might find itself in a spiral galaxy again: but this time in the central bulge, rather than the arms.

The Galaxy and Dark Matter

There are two reasons why the timescale for these events is difficult to estimate. The first is due to limitations on what measurements can be made. As fast-moving trains became common in the nineteenth century, it was noted that the notes from their whistles or sirens seemed to change pitch as the train came toward, or moved away from, the observer. This is the Doppler effect: so-called because it was first investigated by the Austrian scientist Christian Doppler. It was soon realized that this effect occurred for any signal that came in the form of waves, including light. Subsequent analysis of light from stars and other celestial objects showed, indeed, that their spectra were often shifted either to the red or to the blue. The size of the shift gave the speed away from us (for a red shift), or toward us (for a blue shift). The great thing about this method of measuring speed is that it can be applied to any object whose spectrum can be obtained. With the equipment available nowadays, this means that speeds can be estimated out to the most distant reaches of the universe. The drawback is that the Doppler effect only measures movement toward or away from the observer. If the object is moving across the line of sight, there is no Doppler effect. So when a train is actually passing you by, rather than approaching or receding, any sound that comes to you is at its normal pitch. Here is the problem: we know that the Andromeda galaxy is approaching us from measurements of its Doppler shift. But we do not know from these measurements whether it is approaching us head-on or at an angle.

The second limitation has only been realized relatively recently. The gravitational interaction between different bodies depends on their masses. For galaxies in the Local Group, this can be estimated from their sizes and contents. The gravitational pulls derived from these estimates can be used to calculate how the different galaxies in the Group will move in the future. But it has become increasingly evident that this does not work: the estimated mass does not match the gravitational pulls. One piece of evidence for this statement comes from observations of our own Galaxy. As we have seen, stars in the spiral arms are following orbits round the central bulge. We would expect the speeds of these stars to fall off appreciably with increasing distance from the central bulge. This does not happen. The only obvious way of explaining why not is to suppose that some other mass of material, besides the bulge, is affecting the speeds at which the stars move. Such additional material would also help resolve other problems. For example, we have seen that nearby dwarf galaxies are being torn apart by the gravitational effect of our Galaxy. Yet an examination of these galaxies suggests that they are breaking up more slowly than would have been predicted. Again the easiest explanation is to suppose that dwarf galaxies have more mass than we can see, and that this is holding them together.

But what is this invisible material—*dark matter*, as it is usually labeled—which is providing additional mass? To produce the effects we see in the spiral arms, it must be distributed as a kind of halo round the Galaxy (not to be confused with the halo of visible matter that has been discussed above). Moreover, there must be a great deal of it. To explain the rotation of the Galaxy requires it to contain much more dark matter than visible matter. Attempts at explaining how you can have large amounts of invisible material have gone down two paths. The first supposes that it is simply ordinary matter that is not very luminous. For example, brown dwarfs would be difficult to detect even if they existed in large numbers in the halo. Another possibility is a lot of smallish black holes. All suggestions of this sort are lumped together and referred to as *MACHOs* (Massive Compact Halo Objects). The second approach suggests that dark matter is different from ordinary matter, and does not interact with it. The most popular suggestion here is that we are dealing with a new category of particle, the *WIMPs* (Weakly Interacting Massive Particles), though it may be that particles we do know about—neutrinos—are also involved. Neutrinos are called hot dark matter, because they whiz around at nearly the speed of light. WIMPs are cold dark matter: their progress is likely to be more stately. At present, attention is concentrated on WIMPs as the most useful explanation. Whatever dark matter is, its gravitational effect clearly has a significant effect on galaxies. Yet it does not greatly alter the general picture of how galaxies collide. Suppose dark matter consists of something like WIMPs. The dark matter haloes of two colliding galaxies will neither interact directly with each other nor with the ordinary matter in the galaxies. So it continues to be the interaction between the ordinary matter in the two galaxies that defines the future appearance of the merged body. The dark matter has to rearrange itself round the final state, though of course its gravitational pull will affect the details of what happens.

Galaxies and Clusters

Compared with discussing the evolution of stars, discussing the evolution of galaxies has turned out to be a good deal more complicated. It all hinges on whether the galaxy is isolated from other galaxies, or whether it interacts with them. An isolated spiral will fade gently away in terms of its visible appearance. But the more time that passes, the fewer the galaxies that will not have been involved in encounters with other galaxies. How they will be changed depends on what they encounter. For example, if a spiral galaxy encounters only galaxies that have a tenth of its mass or less, it will usually develop like an isolated galaxy. If it encounters another galaxy that has more than a quarter of its mass, it will probably convert into an elliptical galaxy. For galaxies of intermediate mass, what happens depends on the circumstances of the encounter. In discussing the development of galaxies beyond the Local Group, the important factors for their future are therefore their size and what sort of encounters they will have with other galaxies.

If galaxies are interacting strongly, we can hope to spot them because they will have unusual characteristics. Indeed, many of the galaxies classified earlier in this chapter as "peculiar" are found to be interacting with other galaxies. Their shapes can be understood—at least in general terms—as the result of the interaction. For example, there are "ring" galaxies where the central bulge is surrounded by a ring rather than spiral arms. Such a galaxy can result from the collision of two spirals flat-on to each other. Or there are "sacred mushroom" galaxies, which result when one of the colliding galaxies is oriented at right angles to the other. There are various types of "starburst" galaxies, where the interstellar gas that has been compressed by a collision is busy forming a new generation of stars. In some cases, interaction and even merging is occurring between more than two galaxies at the same time. Such major events are easily spotted, but less obvious interaction is even commoner. For example, a detailed search often shows bridges of material joining nearby galaxies together.

The most vigorous interaction between galaxies is often found in the smaller groups of galaxies such as the Local Group. Looking out into the universe, however, the most obvious clusters are the big ones. The nearest such big cluster to us lies in the constellation Virgo. Its center is some 50 million light-years away, and the cluster has over a thousand members (plus small attendants that are difficult to identify at this distance). It also contains a considerable quantity of intergalactic material—most of it torn out, or thrown out, from galaxies in the cluster. Some of this may condense to form new galaxies in the future. Large clusters, such as the Virgo cluster, are held together by quantities of dark matter, both in intergalactic space and in haloes round its members. The galaxies have had time to interact extensively with each other since the cluster was born, and their mutual gravitational pulls have concentrated the galaxies toward the center of the cluster. It might be expected that having a large number of galaxies together in a cluster—especially if they are concentrated toward the middle—would increase the likelihood of collisions between them. In fact, it works the other way. The size of the gravitational pulls means that the galaxies near the center are moving at much higher speeds than the galaxies in smaller clusters. Galaxies near the center of the Virgo cluster hurtle past each other at 1,500 kilometers per second or more—five times greater than the speed at which our Galaxy and the Andromeda galaxy are coming together. Because they swing past each other more rapidly, they have less time to interact. Quick encounters favor galactic harassment—the removal of some fraction of a galaxy's contents—rather than total mergers. Some of the material pulled out goes into intergalactic space, where it can affect other galaxies as they plow their way through it. The stars in these galaxies continue regardless, but any interstellar gas they contain can be pushed about and sometimes lost. The overall result of these various interactions is that galaxies near the center of the Virgo cluster are typically found to be deficient in interstellar gas, limiting their ability to form stars in the future.

At the same time, the central regions of the cluster contain a higher proportion of large elliptical galaxies than the outer regions. The implication is that mergers of spirals do occur, as well as the slower stripping of material. The largest galaxy in the cluster is an elliptical galaxy labeled M87 (meaning that it was the eighty-seventh object in a catalogue put together by the French astronomer Charles Messier in the eighteenth century), which lies close to the center of the whole cluster. This giant elliptical contains as much material as thirty galaxies like our own put together. Its outermost layers are clearly being disturbed by interaction with other nearby galaxies and the incorporation of material from them. M87 has one noticeable peculiarity—a long jet of material shooting out from its center. This is almost certainly connected with a huge black hole—a thousand times more massive than the black hole at the center of our Galaxy—which is believed to exist there. When galaxies merge the dominant influence is the central regions, since this is the most massive part. The odds are that black holes at the centers of the merging galaxies will therefore come together. The interesting bit, however, is the infall of other material into the central black hole. We have seen that when material falls into a black hole it becomes visible, because falling inward compresses it and heats it up. How visible it is depends on the amount of material available: a black hole consuming nothing is, by definition, invisible. A sudden input of material from another galaxy, due to stripping or merging, can cause a rapid flareup, as the new material struggles to reach the black hole. Indeed, the heating can become so severe that some of the material actually shoots outward again. This is the likely explanation of the jet in M87.

Equally, if a galaxy has a central black hole consuming newly input material, it should be brighter at its center than a normal galaxy. As it happens, one of the peculiarities identified in a range of galaxies is that they do have brighter centers than expected.

Such galaxies are collectively referred to as having *active galactic nuclei*. Now if material is falling into the center of a galaxy, the odds are that some will fall into other parts as well. If it falls into regions containing interstellar gas, this will be compressed and new stars will start forming. It is found observationally that galaxies with active galactic nuclei are also quite often starburst galaxies. When our Galaxy and the Andromeda galaxy collide the probability is that, in the transition to an elliptical galaxy, the black holes at the centers of the two galaxies will eventually merge. (Galaxies are already known which have two black holes near the center. In one case, the black holes are only some 25 light-years distant from each other.) The merged galaxy will develop an active galactic nucleus for as long as the fuel supplied by the infalling material lasts. At around the same time, large numbers of stars will be created round the new center. Some of these will come from the mutual compression of interstellar gas in the Andromeda galaxy and our own. But other processes will also be at work. For example, material jostling to get into the black hole will be compressed, and can create its own set of stars. Whatever the processes, the Sun will find itself in a starburst galaxy, as well as one with an active center. Ultimately, all the galaxies of the Local Group should be drawn together into one giant galaxy, so transforming themselves into a "fossil" group. We currently live in a fairly sober spiral galaxy. In due course, this should become a reasonably sober elliptical galaxy. However, the events that happen in between the two should be interesting.

The Virgo Supercluster

A detailed look at the Virgo cluster reveals that its constituent galaxies are not concentrated smoothly toward the center of the cluster. Instead, some of the galaxies are clumped together into subclusters within the main cluster. (M87 is at the heart of one such subcluster.) One possible explanation is that the main cluster, as it has formed, has incorporated smaller clusters into its composition. This raises a question: galaxies can merge together, but can clusters of galaxies also merge? A look at the environment of the Virgo cluster shows that it is surrounded by smaller clusters of galaxies, all of which must come under its massive gravitational pull. The whole group of maybe a hundred clusters is usually referred to as our *Local Supercluster*, though only an astronomer would think of using the word "local" here. It actually stretches for some 160 million light-years. There has not been time since the universe began for galaxies to interact extensively over this sort of distance. As a result, while the Virgo cluster itself looks nice and symmetrical, the various bits of the Supercluster bulge in and out in a fairly irregular way. If we suppose for the moment that the main Virgo cluster is stationary at the center, then its gravitational pull inward will cause satellite clusters to move toward it, and eventually to merge with it. The time needed for the merger of clusters of galaxies is obviously longer than that required for two individual galaxies to merge, being measured in billions of years. The infalling clusters would produce changes in the main cluster—for example, by leading to more collisions between galaxies. Our own Local Group is lurking on the fringes of the Supercluster. Between us and the center of the Supercluster are a series of small groups of galaxies not all that different from our own, such as the Sculptor group (6 million light-years away); the M81 group (10 million light-years); and the M101 group (25 million light-years). But even here on the outskirts we are clearly being attracted toward the Virgo cluster. We are currently moving toward it at the same sort of speed (a few hundred kilometers per second) that we and the Andromeda galaxy are approaching each other. If we get closer to the center of the Virgo cluster, our speed will increase. The problem, as we will see in the next chapter, is whether we can get closer.

9. The Universe

Superclusters

The Virgo cluster with its attached groups of galaxies is far from being the only supercluster in the universe. It has been estimated that there may be over a hundred superclusters within a billion light-years of our own. Our local supercluster is not even the largest. But it does seem to be reasonably characteristic. For example, other superclusters also consist of a large central cluster of galaxies surrounded by an extended halo of smaller groups. As with our own supercluster, these haloes are appreciably flattened. Looking out at neighboring superclusters helps us to understand why.

If the positions of galaxies within a billion light-years of our own are plotted on a map, it becomes evident that superclusters are not distributed at random throughout space. Instead they are embedded in long ribbons of galaxies—perhaps several hundred million light-years long, but only tens of millions of light-years thick. These ribbons are usually called *great walls*, maybe because they remind astronomers of the Great Wall of China winding its way endlessly across the countryside. Between the great walls, there are large voids of space where relatively few galaxies are to be found. Indeed, there are small voids even within the great walls themselves. As a result, the universe looks a bit like a giant sponge with holes of all sizes threading between a network of material. It follows from this picture that superclusters are flattened because they are embedded in ribbon-like structures, which are less wide than they are long. The question, of course, is why does this sort of structure exist? Computer simulations of the past history of the universe—including suitable amounts of dark matter—actually reproduce quite well its sponge-like appearance.

So it seems that the great walls are simply a reflection of the large-scale effects of gravitation over the billions of years since the origin of the universe.

Looking at the Virgo cluster also gives a hint of other things that are happening. We have seen that the cluster has a number of giant elliptical galaxies close to its center. The interesting thing is that they are all elongated so as to point in much the same direction. If this direction is traced across the sky it is found to end up in the center of another supercluster of galaxies (with the rather unromantic name of Abell 1367) 150 million light-years away. Moreover, the nearest supercluster to Abell 1367—the Coma supercluster—seems to be oriented along the same line. The obvious implication is that all three superclusters form part of the same ribbon development. The similar orientation both of clusters and of the galaxies within them suggests, moreover, that both have grown preferentially along the line of the ribbon in which they are embedded. Such orientations are not limited to our own group of

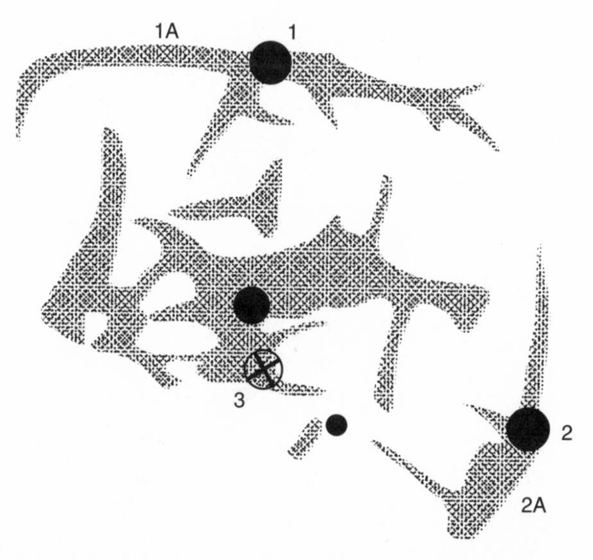

9.1 The large-scale structure of the universe: the shaded areas represent the places where galaxies accumulate most densely. 1 the Coma cluster; 1A the Coma wall; 2 the Perseus cluster; 2A the Cetus wall; 3 the position of our own Local Group with, above it, the Virgo cluster.

superclusters: the same factors seem to be at work in all the ribbons around us.

Since gravitational forces are still acting now, they are still molding the development of clusters and superclusters. We have seen that our Local Group is being attracted to the Virgo cluster at a speed of over 200 kilometers per second. But at the same time it is moving toward another supercluster in the constellation of Centaurus at over twice that speed. The implication seems to be that superclusters may be collapsing inward, but at the same time different superclusters are attracting each other. It might be supposed that the overall result at some distant time in the future must be the collapse inward of our part of the universe. However, the argument is incomplete. We have so far ignored one of the best-established features of the universe—its expansion.

The Expanding Universe

The big telescopes that became available in the first half of the twentieth century made it possible for astronomers to examine the properties of faint galaxies. In particular, they could record their spectra. These could then be measured to determine, via the Doppler effect, whether the galaxies were moving toward us or away. It turned out that the spectra of all galaxies, except for our closest neighbors, showed a shift toward the red end of the spectrum. This meant that they were all moving away from us. Moreover, the amount of the red shift depended on the distance of the galaxy. The more distant the galaxy, the faster it was moving away. This result has been extended to increasingly distant galaxies over the past century: the results leave no doubt that the universe as a whole is expanding. Indeed, the observations make it possible to pin down the rate of expansion fairly accurately. For every million light-years of distance away from us, the speed of a galaxy increases by 22 kilometers per second. (This figure is known as *Hubble's constant* after Edwin Hubble, the American astronomer who first showed that the universe is expanding.) In other words, a galaxy that is a million light-years away will be leaving us at a speed of 22 kilometers per second; one at 2 million light-years will be receding at a speed of 44 kilometers per second; and so on.

The idea of an expanding universe has to be handled with some care. When we talked about our Local Group moving toward the Virgo cluster, we were supposing that space sat still while we moved through it. The expansion of the universe is different from that: space itself is changing. Imagine putting a vast currant cake into the oven, and suppose you are sitting on one of the currants. As the cake rises and expands, the currants move further apart. From your viewpoint, all the currants round about are moving away from you. The further away they are, the faster they seem to be moving. It does not matter which of the currants you are sitting on, this is the impression you will get. Yet the currants are not moving relative to the cake. The bit of the cake that was around you initially will still be around you as the cake expands. It is simply the size of the cake that is changing.

In our universe, the currant corresponds not to an individual galaxy, but to a group of galaxies. We know that the Local Group is held together by its gravitation. The Andromeda galaxy is approaching us, not receding. But what about our local supercluster? The space between members of the supercluster is increasing at the rate determined by the Hubble constant. According to current estimates, the expansion speed at the distance of the center of the Virgo cluster is slightly higher than the speed with which the Local Group is being attracted toward that center. If this is true, we will never catch up and combine with the central cluster. But the uncertainties in the measurements are such that they leave the result on a knife edge. Perhaps we will edge toward the center of the cluster, or perhaps not. However, this conclusion does not allow for more recent developments in our understanding of the universe.

One of the continuing puzzles about the universe has been why it expands at the rate it does. Starting out from a Big Bang, there are three possibilities. In the first, the universe expands relatively slowly. In this case, the gravitational pull backward will make the universe slow down, and ultimately collapse back again. In the second, the universe expands rapidly out toward infinity, and simply keeps on going. The third possibility lies between these. The universe expands out at a middling speed. The expansion rate is just sufficient to prevent a collapse backward, but it takes for ever to reach infinity. (The situation is similar to launching a rocket from the Earth. If its speed is small, it falls back to the Earth. If its speed is high, it will continue out into the solar system. In between there is a speed at which the rocket can just get away.) Obviously, there are many possible low and high speeds, but one unique speed dividing them. The puzzle posed by our universe is that it is expanding at a rate close to this unique speed. Most attempts to explain this surprising result have concentrated on conditions in the first brief instants of the Big Bang. The most popular current explanation is that there was then a period of very rapid inflation. The whole of the presently visible universe was, prior to the inflationary period, contained in a volume that could be held in the hand. This theory actually predicts that the universe should expand at precisely the unique speed. Because this speed depends on the gravitational attraction of the whole universe, the theory therefore also predicts what the present density of material in the universe should be.

We defined dark matter as material which has a gravitational pull, but whose other characteristics are uncertain. All the calculations suggest that there is more dark matter in the universe than there is of the ordinary visible matter with which astronomers usually deal. Even so, adding the dark matter and ordinary matter together does not produce the amount of matter in the universe required by the inflation theory. Recently, it has been suggested that the universe contains an additional component—"dark energy." Just as the nature of dark matter is unclear, so too is the nature of dark energy. But remembering that, as Einstein showed, mass and energy are interchangeable, enough dark energy could certainly contribute toward the amount of material required by inflation theory. Looking out into the universe is an exercise in looking back in time. Telescopes nowadays are capable of obtaining spectra of galaxies billions of light-years distant, which means that we can map the expansion of the universe over several billion years into the past. It had previously been assumed that the expansion always continued at a steady pace. Recent results suggest that this is wrong. Rather, the expansion seems to be accelerating now, as compared with the early days of the universe. Present-day acceleration should not be confused with the very rapid expansion due to inflation in the early Big Bang: it is relatively much slower. But it does require the existence of extra energy to give the universe the required push. This requirement is, indeed, one of the strongest pieces of evidence for dark energy. Clearly, such extra energy will affect the future evolution of the universe—a point to be looked at later. More immediately, any such acceleration will make it more difficult for superclusters to collapse. For the Local Group, this additional expansion will tilt the balance against our ultimate incorporation into the central Virgo cluster.

If the future of our supercluster is in doubt, we can at least discuss the future of smaller groups of galaxies, such as our own Local Group. They should be able to stick together for a good long period of time. Star formation in our part of the universe probably peaked some 6 billion years ago. Most galaxies are now down to a few percent of that peak value. Within the Local Group, there will be a jump in the star formation rate when our own Galaxy merges with the Andromeda galaxy. Similar increases will occur in other groups as their larger galaxies collide. But the increase simply means that their interstellar material will be consumed more rapidly. Though some recycling of material to form new stars will occur, there will be only a limited number of stars forming by several billion years from now. Virtually all the brighter, shorter-lived stars will have disappeared from the scene. Galaxies will then be dominated by red dwarf stars—that is, stars at the bottom end of the main sequence—and will be very dim compared with their peak brightness. But, because red dwarfs consume their hydrogen very slowly, the galaxies will maintain this low light level for a very long period of time. Current estimates suggest that, depending on the galaxy concerned, the red-dwarf population will last for anything up to 100 trillion years (where a trillion is a million million).

When the red dwarfs have finally come to the end of their careers, galaxies will consist mainly of brown dwarfs, white dwarfs, neutron stars and black holes. The first two will be much the commonest objects, and will, by this time, have cooled down very greatly. To an external viewer, a future galaxy will be detectable mainly because it gives out a small amount of heat radiation. Collisions between stars will continue to be rare throughout all these changes in the galaxy. But we are now beginning to deal with very long periods of time. By the time 1,000 trillion years have elapsed, many stars will have been involved in collisions. What happens will depend on which of the stars are involved. If a white dwarf and a brown dwarf collide, the former is so dense that it simple plows through the brown dwarf, blowing it apart (though the white dwarf may carry away some of the debris with it). If two white dwarfs collide, they are likely to merge. The result will depend on how massive they are. The merger of two low-mass white dwarfs should lead to the formation of an ordinary star. The merger of two higher-mass white dwarfs will be more spectacular: the likely result is a supernova. This could happen to our own Sun, which is why it was said at the end of the first chapter that the Sun might "end in fire." The result of a collision between two brown dwarfs also depends on how massive they are. Two low-mass brown dwarfs simply give another brown dwarf, but the merger of two higher-mass brown dwarfs can produce a red dwarf. A snapshot of the galaxy at this period in the future will therefore show light from perhaps a hundred red dwarfs at a time plus very occasional light from an ordinary star or a supernova.

Near misses will, of course, be commoner than actual collisions. The gravitational pulls involved will sometimes act in such a way as to speed up the motion of a star. It may then find itself thrown out of the galaxy altogether. Sometimes, the pulls will act in the opposite way, and throw a star down into the central black hole. So, at the same time as collisions are occurring, the galaxy will be shrinking. Roughly speaking, half of its contents will be lost to intergalactic space and the other half will be swallowed by the central black hole. These changes are likely to be accompanied by a disappearance of the dark matter halo. If this consists of particles—WIMPs—they are expected to interact very slowly with ordinary matter. The denser the matter, the more readily they will interact. A future galaxy with large numbers of white dwarfs can supply the dense material. A WIMP that interacts with a white dwarf will be annihilated (producing a tiny amount of heat in the star), so that, during the period when shrinkage is taking place, a galaxy should also lose much of its dark matter. The loss of the dark matter's gravitational pull will, in turn, make it easier for stars to escape from the galaxy. The overall result is that after some 1,000 trillion trillion years the remaining feature of a galaxy will be its central black hole. By this time, the amount of material it has consumed will have made it supermassive. For a galaxy like our own, it might have a mass equivalent to 100 billion suns.

This is not the end of the process. Where clusters of galaxies still cling together gravitationally—as may happen for the central core of the Virgo cluster—there will now be a cluster of supermassive black holes. As these black holes move round each other, they will emit gravitational radiation. The principle is the same as for the radio or light waves produced when electrically charged particles pass each other, but the gravitational waves are much weaker. Yet, even so, they remove energy. The result, over periods of many million trillion trillion years, is that the cluster of supermassive black holes spiral slowly together till they finally form one huge black hole.

Over even longer periods of time—many trillion trillion trillion years—the remaining objects, now wandering about freely in space, will break up. Protons are usually thought of as very stable particles, but given enough time they may break down, producing a small burst of radiation. From the viewpoint of a white dwarf (by this time, of course, actually a black dwarf), its material will therefore gradually dwindle away, disappearing into space as radiation. The same applies to neutron stars, or to the remnants of brown dwarfs, or to any other material outside black holes. Even black holes are not stable for ever. Over these huge time spans, particles slowly evaporate from black holes. The more massive the black hole, the longer it takes; but ultimately, almost innumerable trillions of years in the future, even the huge black holes formed from clusters of galaxies will evaporate away. The final view of the universe is therefore of nothing much but radiation plus some very stable particles (such as electrons). The whole will be extremely rarefied, yet getting still more dilute as the universe continues to expand. The effects of dark energy will underline this isolation. Within a relatively moderate timescale of the order of 100 billion years—when our Galaxy may still contain stars with planetary systems—nothing will be visible to any intelligent life form that then exists except a single, greatly enlarged supercluster. By the later stage, when our universe consists only of radiation and particles, expansion will have reached the point where no individual particle or photon of radiation will ever encounter another. In T.S. Eliot's words: "This is the way the world ends, not with a bang but a whimper."

Life in the Universe

And what of life in all this? Presuming that life requires conditions similar to those found on Earth, we can try to estimate how many stars can produce a suitable habitat. As we have seen, life requires a stable platform—the surface of the planet—on which to develop, and depends on the presence of liquid water. Planets have been detected round a number of stars in recent years. As yet, they are larger planets like Jupiter, though there are hopes of detecting Earth-sized planets in the near future. However, the observations indicate that not all planetary systems are suitable for life. The need for liquid water restricts planets with life to a limited region—the habitable zone—round the central star. For some stars, this region is either occupied by large planets, or the planets have elongated orbits. In neither case would an Earth-like planet find it easy to follow a stable orbit that could nurture life over a long period of time.

The Sun was born at the end of the main star-forming period in the Galaxy, but even now stars similar to the Sun are being born at the rate of a few thousand a year. The starburst episode when our Galaxy and the Andromeda galaxy merge in a few billion years will up this rate very considerably. It seems we are somewhere near the middle of a period of 15 billion years when the formation of Sun-like stars is common. After that, the formation of Sun-like stars will die away. Those already existing will evolve away from the main sequence. As we have seen, the changes during this period are likely to kill off any life in the previously habitable zone. But this is not necessarily the end of things. Though stars smaller than the Sun are much less likely to have habitable planets, there are so many of them that a few should have suitable planets. Red dwarfs last for a very long time, so life round such stars could continue into the distant future, when such stars dominate the appearance of our merged galaxy. The main question is whether planets can continue in stable orbits over this period of time. It seems that stars are far more likely to have several planets circling them than to have just a single planet. The gravitational pulls of these planets on each other will ultimately lead to changes in their orbits, so that planets supporting life will wander outside their habitable zone. There is also an ultimate cutoff. One of the results of Einstein's ideas on gravitation is that a planet moving round a central star must radiate waves of gravitation. The consequence is that, very slowly, the planet spirals into the star. The end will come some 100 million trillion years in the future when all planets should have been consumed. What applies to our own Galaxy applies to galaxies in general. All should contain occasional planets that can harbor life, but the likelihood of suitable planets existing will decrease as time passes. The long-term scenario, with everything in the universe gradually dying, is obviously hostile to life.

This description of the future applies to the simplest life forms that can exist. But what about intelligent life? Complex forms of life require more from their environment than simple forms. So fewer planets are likely to have the necessary characteristics. Besides, not all planets that develop complex life forms need necessarily go on to produce intelligent life. At the same time, once intelligent life has appeared, it can try to modify its planet to create an environment suitable for its own continued existence. In terms of looking to our own future, human beings can now avoid normal evolutionary pressures, but will they develop ones of their own? For example, will developments in computing, miniaturization and surgical implants lead to the appearance of mixed machine–human beings—sometimes called "cyborgs" in science fiction? The problem comes in guessing how intelligent life will cope with long periods of time. Civilizations on Earth can be measured in terms of thousands of years. This is totally insignificant on an astronomical timescale. We have very little guidance as to how our society will develop over the next few thousand years, let alone over the Earth's future billions of years of existence. Fermi's paradox—"Where are they all?"—has to be answered if we want to understand the likely development of our own society.

Some people have deduced from the apparent lack of communications from other intelligent beings that civilization is a basically unstable state. On this picture, complex societies break down on a short timescale—though they may reappear again on the same planet at intervals in the future. But this picture is far from the only possibility. Advanced forms of life elsewhere, if they exist, may not use the same sort of technology that we use for communication. Even if they do, unless intelligent life is very common in our Galaxy and sends many signals into space, such communications will remain difficult to detect. Even where intelligent life does appear and forms a stable community, it may well send easily detectable signals into space for only a limited time. The probability that life on another planet in our part of the Galaxy is at exactly the same stage of development as we are is small; so, therefore, is the probability of us picking up recognizable signals. Given the huge number of acceptable stars in the universe, it is unlikely that we are the only intelligent group to have appeared, or which will appear in the future. But equally, given the vast distances and times involved, making contact with others will always be difficult. Actual space travel—as we saw at the beginning of this book—has enormous problems. They are greatly increased when we allow for the fact that the universe is expanding. Perhaps it does not matter. There remains a nagging suspicion that the requirements for intelligent life may prove to be even stricter than we have supposed. If so, the future is unlikely to provide better environments than those we have today. Maybe the last human beings on Earth will find that they have to end with the traditional toast: "Here's to us, who's like us—very few, and they're all dead."

But Have We Got It Right?

All this assumes that our current picture of the universe is more or less correct. There are good reasons for hedging our bets on this one. Recent estimates suggest that the universe consists of about 5 percent ordinary matter, 25 percent dark matter and 70 percent dark energy. Yet all we know about the universe comes from observations of ordinary matter. It is, to say the least, worrying that we apparently cannot observe 95 percent of the contents of the universe. This lack of knowledge may greatly affect our understanding of how the universe will develop in the future. For example, the exact importance of dark energy in the universe is still unclear. It has been suggested that its effects may be much more significant than the previous discussion supposes—to the extent that dark energy might actually rip objects apart. One calculation indicates that a more powerful dark energy could disrupt our future merged galaxy in about 20 billion years from now. Not long after that, it would rip apart individual stars and planets, and then it would shred the atoms themselves. The final stage—particles plus radiation—would not be greatly different from our previous picture, but it would be reached far more quickly. But there is another, though less likely, scenario. Dark energy could, instead of increasing, diminish as time passes. It could even, eventually, begin to attract, rather than repel. Were this to happen, the universe would begin to contract, rather than expand, and the end would then be a "big squelch"—a highly compressed blob, not too different from the original Big Bang.

Nor is dark energy the only area of uncertainty. Ideas about the nature of the universe are currently in a state of rapid change. New theoretical approaches are producing new pictures of the universe, and so leading to differing scenarios of what might happen in the future. One whole set of queries under examination are of the "what if?" type—in this case, what if some of the basic assumptions in our calculations are wrong? Suppose, for example, that light does not remain the same however far it travels, but loses its energy bit by bit as it travels further and further. Remembering that blue light has more energy than red light, this means that light reaching us from distant parts of the universe should be redder than light from nearby objects. But this is just what we see when we look at the spectra of distant galaxies. In other words, if we suppose that light can in some way become "tired" as it travels, we have an explanation of the red shift that does not require the universe to be expanding. The possibility has been tossed about for some years, but laboratory experiments suggest that any such effect must be small—too small to explain the red shift of galaxies. Another thing we might not have quite right is the size of the gravitational pull between bodies. We think we know this very accurately, but suppose the law governing gravitational attraction actually varies a little with distance, so that the attraction between two distant bodies is not quite what we would expect from measurements of two bodies that are close together. A small change of this sort could actually explain current observations of the expansion of the universe without needing to invoke dark matter. Again, the experimental data put quite tight limits on how much deviation from the accepted law of gravitation is permissible, but, in this case, they are not quite accurate enough to rule a slight deviation out altogether. We can, if we wish, go the whole hog, and suppose that all the basic data used in our calculations—what are usually referred to as the *fundamental constants*—may vary a little with time. The list of such constants includes not just the speed of light and the gravitational force, but things such as the electrical charge on an electron, which lie at the basis of our understanding of atoms and nuclei. If all these different factors are varying together, it becomes quite complex not only to establish what the result of such changes will be, but also to set a limit on how much alteration of this type is permitted by the uncertainties in the experimental data. The question has become a matter of debate in recent years, because there have been claims that a variation of the fundamental constants might be permitted by some of the cosmological theories currently being explored. The verdict at present on this proposal can only be "not proven." Though the existence of such changes is not entirely ruled out, it has yet to be demonstrated convincingly.

One of the reasons for this interest in the fundamental constants is because the values they have determine the sort of universe we live in. For example, we believe that life is only possible because the carbon atom has very specific properties. These properties allow carbon atoms to hook themselves together in lengthy chains, so forming the complex molecules required for life. A slight change in the properties of carbon, and we would not exist. It is possible to write down a whole list of things like this that are necessary for life. Another example is the lifetime of the universe. It has taken several billion years for really complex life forms to appear on Earth. A universe that existed for only a billion years (say) would be unlikely to support life. Observations of this sort have led to a lengthy discussion of the *anthropic principle*: the belief that the universe we live in is actually fine-tuned for the production of life. If this is so, then the fundamental constants cannot change too much without endangering the fine-tuning.

Many astronomers feel unhappy with this sort of argument. Why should the universe care whether it is good for life or not? A joke question sometimes set in astronomy examinations is: define the universe and give two examples. The joke, of course, hinges on the fact that the universe is defined as containing everything that exists. However, the joke is no longer so obvious: the question of other universes existing is now a hot topic in astronomy. If they do, one spinoff would be an explanation of the anthropic principle. We could say that many universes exist, but most do not have the basic properties required for life. We are here simply because our universe happens to be one of the minority that does have those properties. Fair enough, but explaining the anthropic principle is hardly a sufficient argument by itself for the existence of other universes. However, other lines of investigation currently seem to be hinting at the same thing. A major goal of theoreticians for many years past has been to find a "theory of everything." Their aim is not quite as all-embracing as this grandiose title might suggest. In essence, they are looking for a theory that will provide an understanding both of the very small (meaning the basic particles, such as electrons) and of the very big (the nature of the universe as a whole). In recent years, new ways of looking at particles have indeed led to new ways of looking at the universe.

It started with string theory. The traditional assumption had always been that the fundamental particles could be treated as points. An important breakthrough came with the suggestion that, instead, they should be thought of as tiny pieces of string. This meant that they were lines, rather than points. Like any piece of string, they could come in various forms. For example, the two ends of a straight piece of string might join together to form a circle. This sounds easy enough to understand. The problem is that, in order for the theory to work, the four dimensions in which we human beings operate—three directions in space plus one in time—proved to be inadequate. Instead, it seemed that ten dimensions were necessary. All of these dimensions, other than our usual four, were tied up so tightly with the strings that they were too small to be noticed. (Though nobody seems to have explained what we might expect to see if any of these extra dimensions became visible.) Now it was necessary to tie this new theory of particles in with our understanding of the universe. Particles and the universe were mostly intimately linked together in the Big Bang. This enormously hot, dense fireball led to the creation both of the particles that we have today and of our present universe. But the application of string theory to the Big Bang did not produce the insights that had been hoped for. It required, eventually, a major change for things to fit together. The important step forward does not actually sound particularly striking to most people. It consisted of shifting from a theory dealing with ten dimensions to one dealing with eleven dimensions. This apparently small alteration led to an entirely new picture of our universe. The tiny strings that theory had previously imagined became stretched out into a vast membrane which embraced all the matter in the universe. The new picture—labeled M theory (M for membrane)—implied something further. There could be other membranes out there like our own, each containing a separate universe. Rather than calling each such universe a "membrane," they have been labeled "branes" (on the somewhat spurious ground that we usually think of membranes as having two dimensions, rather than eleven). There is a widespread feeling that an infinite number of such branes may exist. So M theory leads us back to the idea of a "multiverse."

Although the branes are separate from each other, it seems that some interaction between them might be possible. One suggestion—which goes back to before string theory was invented—

relates to the strange world of the very small. Quantum theory tells us that empty space is not really "empty." Rather it is a vast sea of energy that, for the most part, remains passive and invisible. Particles come into being and almost immediately disappear. The multiverse concept helps here; we can suppose that what we are seeing is simply energy bubbling in and out of our universe from other universes. Another question that a multiverse might solve is a long-standing problem concerning gravitation. Gravitation is much weaker than the other basic forces that contribute to our universe. Hang two small balls close together on pieces of thread. The gravitational attraction between the two is so small that they stay where they are. Put a positive electrical charge on one ball and a negative charge on the other, and the two will immediately swing together. So the question is: why is gravity so weak compared with (say) electrical forces? One interpretation of M theory suggests that gravitation can actually seep between different branes. Consequently, how strong it is in our universe may depend, in part, on what other branes are near our own.

But what does this new approach say about our main interest—the future of the universe? M theory has, of course, been applied to that basic theory-of-everything problem—what happened at the Big Bang? And here it does provide some interesting insights. It is generally supposed that the various branes are not sitting still, but moving about. As a result, they occasionally collide with each other. One suggestion is that the Big Bang was triggered off by just such a collision between our own brane and a neighboring brane. The force of the collision was sufficient to start off the expansion we see today. Indeed, it could have been large enough to have kicked in the "dark energy" component as well. If we accept this interpretation, it does lead to a possible new vision for the future. What has happened once, can happen again. Suppose our universe in old age once more bumps into another universe, what will happen? Of course no-one knows for sure, but it would certainly shake our universe out of its eternal sleep into something new and different. Even if no collision occurs, the future is not entirely dreary. As we have seen, the quantum sea that fills all space is perpetually subject to small fluctuations. Very rarely, such a fluctuation may be large enough for a flurry of particles to appear suddenly, apparently out of nothing. Given an eternity of time, fluctuations in the quantum background can occur on any scale. It is possible that one such fluctuation might be so large that it would set off a whole new expanding universe. Unfortunately, only an eternal being is likely to be around long enough to witness the event.

Further Reading

The books and articles mentioned here should be fairly readily available—for example, on loan or in university libraries. Some of the books are collections of articles by different authors. Not all of the articles in these provide easy reading, but others in the same volume are simpler. Though the entries are grouped under chapter headings, a number of the books provide information that is relevant to more than one chapter. For example, the *Encyclopedia of the solar system* listed under Chapter 6 is also relevant to all the preceding five chapters.

Much astronomical information nowadays is, of course, available online. NASA, in particular, provides a vast amount of material. Their main site can be found at www.nasa.gov. Professional astronomers turn to the NASA ADS Astronomy Abstract Service at adswww.harvard.edu, with a mirror site at ukads.nottingham.ac.uk, which contains information (often including summaries) of all the articles published in astronomical journals, along with some books. A useful site for keeping up with new observations in astronomy and space science is spaceflightnow.com. The various national astronomical societies have sites of their own, which provide both information and links to other useful sites. For example, the American Astronomical Society can be found at www.aas.org, while the site of the Royal Astronomical Society in the UK is at www.ras.org.uk. In addition, many sites provide material intended for reference or teaching. To take three different examples—the Lunar and Planetary Institute at www.lpi.usra.edu covers the solar system; there is student astronomical information at seds.lpl.arizona.edu; and a general directory with good astronomical links can be found at dmoz.org. A search engine, such as Google, will bring up many more.

Chapter 1

B.W. Carroll and D.A. Ostlie, *An introduction to modern astrophysics* (Addison-Wesley, Reading; 1996).

A.N. Cox, W.C. Livingston, and M.S. Matthews (eds.), *Solar interior and atmosphere* (University of Arizona Press, Tucson; 1991).

C.P. Sonett, M.S. Giampapa, and M.S. Matthews (eds.), *The Sun in time* (University of Arizona Press, Tucson; 1991).

R.J. Tayler, *The Sun as a star* (Cambridge University Press, Cambridge; 1997).

Chapter 2

J.B. Murphy and R.D. Nance, How do supercontinents assemble? *American Scientist* Vol. 92, pp. 324–333 (2004).

R.D. Nance, T.R. Worsley, and J.B. Moody, The supercontinent cycle. *Scientific American* Vol. 259, pp. 44–51 (July 1988).

B.J. Skinner and S.C. Porter, *The dynamic Earth* (John Wiley, Chichester; 2000).

Chapter 3

C.M. Goodess, J.P. Palutikof, and T.D. Davies, *The nature and causes of climate change* (Belhaven Press, London; 1992).

T.E. Graedel and P.J. Crutzen, *Atmospheric change* (W.H. Freeman, New York; 1993).

B. Montesinos, A. Gimenez, and E.F. Guinan (eds.), *The evolving Sun and its influence on planetary environments* (Astronomical Society of the Pacific, San Francisco; 2002).

P.D. Ward and D. Brownlee, *The life and death of planet Earth* (Times Books, New York; 2002).

G.E. Williams, History of the Earth's obliquity. *Earth Science Reviews* Vol. 34, pp. 1–46 (1993).

Chapter 4

J. Bloxham and D. Gubbins, The evolution of the Earth's magnetic field. *Scientific American* Vol. 261, pp. 30–37 (December 1989).

J.W. Geissman, Geomagnetic flip. *Physics World* Vol. 17, pp. 31–35 (2004).

G.A. Glatzmaier and P. Olsen, Probing the geodynamo. *Scientific American* Vol. 292, pp. 33–39 (April 2005).

E. Nesme-Ribes, S.L. Baluinas, and D. Sokoloff, The stellar dynamo. *Scientific American* Vol. 275, pp. 30–36 (August 1996).

P.R. Wilson, *Solar and stellar activity cycles* (Cambridge University Press, Cambridge; 1994).

Chapter 5

L. Becker, Repeated blows. *Scientific American* Vol. 286, pp. 62–69 (March 2002).

D.A. King and D.D. Dyrda, The day the world burned. *Scientific American* Vol. 289, pp. 70–77 (December 2003).

M.Ya. Marov and H. Rickman (eds.), *Collisional processes in the solar system* (Kluwer, Dordrecht; 2001).

P.R. Weissman, The Oort cloud. *Scientific American* Vol. 279, pp. 62–67 (September 1998).

Chapter 6

J.K. Beatty, C.C. Petersen, and A. Chaikin, *The new solar system* (Cambridge University Press, Cambridge; 1999).

S. Ferraz-Mello, *Chaos, resonance and collective dynamical phenomena in the solar system* (Kluwer, Dordrecht; 1992).

S.R. Taylor, *Solar system evolution* (Cambridge University Press, Cambridge; 1992).

P.R. Weissman, L.-A. McFadden, and T. Johnson (eds.), *Encyclopedia of the solar system* (Academic Press, London; 1999).

Chapter 7

C. Chiappini, The formation and evolution of the Milky Way. *American Scientist* Vol. 89, pp. 506–515 (2001).

G. Gonzalez, D. Brownlee, and P.D. Ward, Refuges for life in a hostile universe. *Scientific American* Vol. 285, pp. 52–59 (October 2001).

N. Henbest and H. Couper, *The guide to the Galaxy* (Cambridge University Press, Cambridge; 1994).

F. Matteucci, *The chemical evolution of the Galaxy* (Kluwer, Dordrecht; 2001).

F. Matteucci and F. Giovannelli (eds.), *The evolution of the Milky Way* (Kluwer, Dordrecht; 2000).

R.J. Reynolds, The gas between the stars. *Scientific American* Vol. 286, pp. 32–41 (January 2002).

R. Smoluchowski, J.N. Bahcall, and M.S. Matthews (eds.), *The Galaxy and the solar system* (University of Arizona Press, Tucson; 1986).

B.P. Walker and P. Richter, Our growing, breathing Galaxy. *Scientific American* Vol. 290, pp. 28–37 (January 2004).

Chapter 8

G. Byrd, M. Valtonen, M. McCall, and K. Innanen (eds.), *Back to the Galaxy* (American Institute of Physics, New York; 1993).

R.S. Ellis, R.G. Abraham, J. Brinchmann, and F. Menanteau, The story of galaxy evolution in full colour. *Astronomy & Geophysics* Vol. 41, pp. 2.10–2.16 (2000).

G. Kauffman and F. van den Bosch, The life cycle of galaxies. *Scientific American* Vol. 286, pp. 36–45 (June 2002).

G. Lake, Cosmology of the Local Group. *Sky & Telescope* Vol. 84, pp. 613–619 (1992).

C. Struck, Galaxy collisions. *Physics Reports* Vol. 321, pp. 1–137 (1999).

S. van den Bergh, *The galaxies of the Local Group* (Cambridge University Press, Cambridge; 1999).

K. Weaver, The galactic odd couple. *Scientific American* Vol. 289, pp. 26–33 (July 2003).

Chapter 9

F. Adams and G. Laughlin, *The five ages of the universe* (Touchstone, New York; 2000).

J.D. Barrow and J.K. Webb, Inconstant constants. *Scientific American* Vol. 292, pp. 32–39 (June 2005).

A. Dekel and J.P. Ostriker (eds.), *Formation of structure in the universe* (Cambridge University Press, Cambridge; 1999).

A. Fairall, *Large-scale structures in the universe* (John Wiley, Chichester; 1998).

C.S. Frenk, G.E. Kalmus, N.J.T. Smith, and S.D.M. White (eds.), The search for dark matter and dark energy in the universe. *Philosophical Transactions of the Royal Society A* Vol. 361, pp. 2425–2627 (2003).

A.G. Riess and M.S. Turner, From slowdown to speedup. *Scientific American* Vol. 290, pp. 50–55 (February 2004).

M. Tegmark, Parallel universes. *Scientific American* Vol. 289, pp. 30–41 (May 2003).

P. Ulmschneider, *Intelligent life in the universe* (Springer, Berlin; 2004).

Index

Printed in the United States of America.